Introduction to Laser Fusion

Laser Science and Technology
An International Handbook

Editors in Chief

V.S. LETOKHOV, *Institute of Spectroscopy, USSR Academy of Sciences, 142092 Moscow Region, Troitsk, USSR*
C.V. SHANK, *Director, Lawrence Berkeley Laboratory, University of California, Berkeley, CA 94720, USA*
Y.R. SHEN, *Department of Physics, University of California, Berkeley, CA 94720, USA*
H. WALTHER, *Max-Planck-Institut für Quantenoptik und Sektion Physik, Universität München, D-8046 Garching, Germany*

This book is part of a series. The publisher will accept continuation orders which may be cancelled at any time and which provide for automatic billing and shipping of each title in the series upon publication. Please write for details.

Introduction to Laser Fusion

C. Yamanaka

*Director, Institute for Laser Technology,
Osaka, Japan*

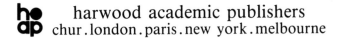

harwood academic publishers
chur.london.paris.new york.melbourne

411935⁄₀

PHYSICS

© 1991 by Harwood Academic Publishers GmbH, Poststrasse 22, 7000 Chur, Switzerland. All rights reserved.

Harwood Academic Publishers

Post Office Box 197
London WC2E 9PX
United Kingdom

Post Office Box 786
Cooper Station
New York, New York 10276
United States of America

58, rue Lhomond
75005 Paris
France

Private Bag 8
Camberwell, Victoria 3124
Australia

Library of Congress Cataloging-in-Publication Data
Yamanaka, Chiyoe, 1923–
   Introduction to laser fusion/C. Yamanaka.
     p.   cm. − (Laser science and technology; v. 10)
   Includes bibliographical references and index.
   ISBN 3-7186-5063-0
   1. Laser fusion. I. Title. II. Series.
   QC791.73.Y36   1990
   621.36′6−dc20
                               90-49479
                                  CIP

Contents

QC791
.73
Y36
1991
PHYS

Introduction to the Series

Almost 30 years have passed since the laser was invented; nevertheless, the fields of lasers and laser applications are far from being exhausted. On the contrary, during the last few years they have been developing faster than ever. In particular, various laser systems have reached a state of maturity such that more and more applications are seen suffusing fields of science and technology, ranging from fundamental physics to materials processing and medicine. The rapid development and large variety of these applications call for quick and concise information on the latest achievements; this is especially important for the rapidly growing interdisciplinary areas.

The aim of "Laser Science and Technology — An International Handbook" is to provide information quickly on current as well as promising developments in lasers. It consists of a series of self-contained tracts and handbooks pertinent to laser science and technology. Each tract starts with a basic introduction and goes as far as the most advanced results. Each should be useful to researchers looking for concise information about a particular endeavor, to engineers who would like to understand the basic facts of the laser applications in their respective occupations, and finally to graduate students seeking an introduction into the field they are preparing to engage in.

When a sufficient number of tracts devoted to a specific field have been published, authors will update and cross-reference their pages for publication as a volume of the handbook.

All the authors and section editors are outstanding scientists who have done pioneering work in their particular field.

V.S. Letokhov
C.V. Shank
Y.R. Shen
H. Walther

Preface

Fusion energy is the most abundant energy source in the universe. It is the source of energy in the sun and stars. Fusion reactions will sustain the sun for billions of years to come. This is the source of all life and activity on Earth. The fusion reaction yields eight times more energy than the nuclear fission of uranium and over a million times more energy than the fossil fuels per unit mass.

Fusion is the combining of nuclei of hydrogen isotopes to form helium. The waste product of a fusion reactor, helium is a non-radioactive, non-toxic and noble gas. Hydrogen bombs are the fusion devices liberating the energy in uncontrolled form. A controlled fusion device for the civil use of fusion energy is earnestly expected to be created. The development for controlled fusion reaction is one of the biggest challenges to mankind. Even if the artificial fusion reaction to produce a large amount of energy is successful, a number of technological and economical problems still must be overcome.

The enthusiastic research and development aimed toward next century's fusion reactor is, however, performing today. There are two approaches-magnetic confinement fusion and inertial confinement fusion. Inertial confinement provides high density plasma more than $10^{25}/cm^3$ for a confinement time of 10^{-10} sec, so that the reaction can proceed extremely quickly as a microexplosion. The compression is achieved by the irradiation of laser light or by a flux of energetic particles on a fuel shell. The fuel target is heated to vaporize the surface which ablates off at high velocity. This produces a strong recoil force which compress the fuel. When the thermonuclear conditions are reached (1000 times liquid density $\sim 200gr/cm^3$ and 100 million degrees centigrade), the core will ignite. While inertial forces hold the compressed fuel, the fusion reaction propagates outward, releasing much more energy than was used to compress it. Fusion reactors require high energy gains from repetitive burns.

The key issues of current conceptual designs of inertial fusion reactors are to have modest dimensions, effective liquid metal heat transport loops, and the structural isolation of the driver from the reactor itself. These engineering features are very important for the development of inertial fusion as a new energy source which is not at all inferior to magnetic fusion schemes.

This tract as the title suggests, serves as an introduction to laser fusion. It covers a whole range of ICF which extends from fusion reactions to the laser interaction with plasmas and implosion dynamics. The properties of various drivers are reviewed and the present experimental status is introduced. And finally the fusion reactor concepts and future prospect are briefly described.

Several tracts concerning these introductory topics in depth will be published as part of the Laser Science and Technology Series.

The author wishes to acknowledge the contributions of his colleagues, especially Dr. T. Jitsuno for chapter 3 and chapter 8, and Miss. J. Horimoto for her help in preparing the manuscript.

Chiyoe Yamanaka
Osaka

1 INTRODUCTION

After the invention of the laser in 1960, programs for laser fusion research were soon proposed in the USA and USSR. The experiments were going on by focusing laser beams to a target under classification. In 1964, Basov *et al.* reported a neutron yield from a deuterated lithium target irradiated by a glass laser. E. Teller proposed a concept of a new internal combustion engine which was based on the implosion of DT fuel in 1972 (Nuckolls et al.). This was a very interesting idea to reduce a necessary driver energy inversely proportional to the square of the fuel density.

When the deuterium-tritium fuel is compressed to 1000 times of the liquid density, $200 \, g/cm^3$, Teller's proposal demands only $10 \, kJ$ of the energy deposition on a fuel by laser. Then the enthusiasm and faith of scientists had initiated a large research effort in the inertial confinement fusion field. As lasers can concentrate a large amount of energy on a fuel pellet in an extremely short time, several programs in the USA, Japan, USSR and Europe are pursued. For them, the key issues are the development of high power lasers, a suitable fuel pellet fabrication and design, pertinent diagnostics of plasmas and the computer simulation technology.

By recent investigations, a scaling law of the implosion demands a laser of $100 \, kJ$ in blue for ignition, $1 \, MJ$ for a breakeven condition and several MJ for a reactor of the high pellet gain. These are in the present art of technology.

Research on laser fusion has produced several spin off effects at the frontier of science and technology. The high power laser has been developed due to needs of fusion research. Not only the optical technology but also the pulse power technology, sensing technology, tritium handling, computer simulation methods have made great progress in inertial confinement fusion research.

The main theme of laser fusion is so deeply concerned with the high temperatures and ultra dense plasma that the astronomical interests are excited. The shock phenomena, X-ray radiation, and the magnetic field generation are involved with laser produced plasmas. These basic issues are also very important keys in the research of laser fusion.

1

2 FUSION RESEARCH

The cumulative demand for energy in the world is estimated to be 11Q (Q = 10^{21} Joule) during the period of 1950 to 2000 (Gross, 1984). This will increase to 61Q for the next fifty year period of 2000 to 2050. The total world supply of fossil fuels is estimated as 30 ~ 50Q. Nuclear fission can supply 3Q by ^{235}U and 200Q by breeding other fissionable fuel. Unfortunately, fossil fuels lead to environmental hazard such as air pollution and also the greenhouse effect due to the contamination of carbonoxide gas in the atmosphere. Fission reactors produce radioactive wastes which are very critical issues of the future. Therefore, an abundant and clean energy resource, nuclear fusion will be very essential to man in the next century. Table 2.1 shows energy resources on the earth.

Fusion reaction is to join light nuclei forming a new nucleus where mass defect Δm changes to energy W by Einstein's relation.

$$W = \Delta m \cdot c^2$$

where c is the velocity of light and $\triangle$m is the difference between the masses of light nuclei and the mass of the product nucleus.

For such a reaction to take place, the particles must come within range of the nuclear forces which means that the Coulomb barrier must be overcome by the kinetic energy of the colliding particles. To get energy gain, one should have a high density, high temperature plasma for a

Table 2.1 Various energy resources on the earth in Q unit.

Resources	Known resources		Estimated resources
fossil fuel	23		452
fission fuel	cost($/lb)		cost($/lb)
	5~10	<500	<500
	490	1.6×10^6	5×10^6
fusion fuel	7.5×10^9		7.5×10^9

$$1Q = 10^{18} BTU = 1.055 \times 10^{21} \text{ Joule}$$

3

certain reaction time. At a given temperature, the reaction rate is proportional to the square of the density.

The most attainable reaction is

$$D + T \to {}^4He(3.5\,MeV) + n(14.1\,MeV)$$

In this case, the product of reaction time τ and the density n is $10^{14}\,sec/cm^3$ at a temperature of 10^8 °C. The generated energy is divided in 80%, 14.1 MeV to the neutron and in 20%, 3.5 MeV to the α particle. Therefore, a direct energy conversion to electricity by charged particles is not effective. Generated energy will be used as a heat source.

In this D-T reaction, the 14 MeV neutron has such high energy that the first wall of the reactor is severely damaged.

As shown in Table 2.2, the D-D reaction has a smaller reaction cross-section which needs a higher $n\tau$ value, $n\tau \sim 10^{16}\,sec/cm^3$ at the temperature of 2×10^8 °C. The generated energy is almost equally given to the neutron and the charged particles. A fusion reaction with no neutron yield is very preferable, because there is no damage to the structural materials and no radioactive effect in the reactor vessel. P-^{11}B reaction is one of these candidates, however it demands $n\tau \sim 10^{16}\,sec/cm^3$ at 10^9 °C. This reaction condition seems to be harder than that of the hydrogen isotopes.

The Figure 2.1 shows reaction cross-sections of various nuclei at different temperatures. The D-T reaction is most promising to attain.

As for a necessary condition to produce nuclear fusion energy, the generated energy should be larger than the thermal energy given to the plasma fuel,

$$\left(\frac{n}{2}\right)^2 (\sigma v)\omega\tau > 2 \cdot \frac{3}{2}\, kTn$$

where n is the plasma density, σ is the fusion reaction cross-section, v is the plasma thermal velocity, τ is the reaction time, ω is the generated energy per one reaction, k is the Boltzmann constant and T is the plasma temperature. The left hand side of this equation is the generated fusion energy per τ sec and the right hand side is the plasma energy. To produce an effective D-T reaction, the plasma temperature must be 10^8 K to give the necessary colliding velocity of nuclei to overcome the Coulomb repulsion. This velocity reaches 300 km/sec. When the above equation is used with these data, it becomes

$$n\tau > 10^{14}\,sec/cm^3$$

This is the well known Lawson's criterion. In the magnetic confinement fusion, the plasma density is restricted to $10^{14}/cm^3$ by the limitation of the magnetic field and the confinement time is as long as 1 sec. For the

Table 2.2 Fusion reaction.

1) D-T cycle

$$D+T \rightarrow n+\alpha+17.6\,\mathrm{MeV}$$
$$n+{}^6Li \rightarrow T+\alpha+4.8\,\mathrm{MeV}$$
$$n+{}^7Li \rightarrow T+\alpha+n'-2.6\,\mathrm{MeV}$$

2) Catalyzed D-D cycle

(I)
$$D+D \rightarrow n+{}^3He+3.27\,\mathrm{MeV}$$
$$\rightarrow p+T+4.03\,\mathrm{MeV}$$
$$D+T \rightarrow n+\alpha+17.6\,\mathrm{MeV}$$
$$D+{}^3He \rightarrow p+\alpha+18.4\,\mathrm{MeV}$$

(II)
$$D+D \rightarrow n+{}^3He+3.27\,\mathrm{MeV}$$
$$\rightarrow p+T+4.03\,\mathrm{MeV}$$
$$T \rightarrow {}^3He+\beta^-$$

3) p ^{11}B cycle

$$p+{}^{11}B \rightarrow 3\alpha+8.7\,\mathrm{MeV}$$

4) p ^{6}Li cycle

$$p+{}^6Li \rightarrow {}^3He+\alpha+4.02\,\mathrm{MeV}$$

inertial confinement fusion, the plasma density of a solid deuterium is $10^{22}/cm^3$, then the confinement time is 10^{-8} sec. This is a comparable time duration of the laser pulse.

When the laser light irradiates a fuel pellet it turns to a plasma. The expansion motion of the plasma ball is limited by the ambipolar diffusion which means the run away electrons are pulled by the electric field due to the ion charges which remain in the core of the plasma. Then the expansion velocity is determined by the velocity of ions. This speed S is about 10^8 cm/sec, the sound velocity of plasma at 10^8 K.

Putting the radius of a pellet equal to R, the confinement time τ is R/S, then 10^{-8} R, Lawson's criterion $n\tau = 10^{14}$ sec/cm³ becomes nR = $10^{22}/cm^2$. If we use the weight density ρ for the number density n,

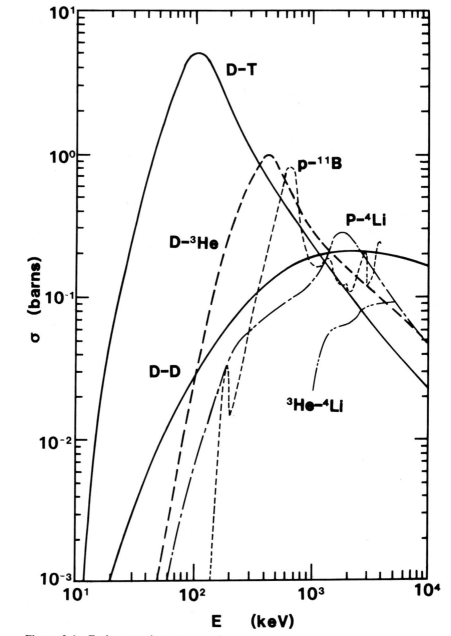

Figure 2.1 Fusion reaction cross section.

the inertial confinement criterion is $\rho R = 0.1\,\text{g/cm}^2$. We can estimate the necessary laser energy E_L for the inertial confinement fusion as,

$$E_L = \frac{4}{3}\,\pi R^3 \cdot 3nkT \cdot \frac{1}{\varepsilon}\ [J]$$

$$= \frac{4}{3}\,\pi \frac{(\rho R)^3}{\rho^2} \times 4.6 \times 10^8 \times \frac{1}{\varepsilon}\,[J]$$

where ϵ is the laser plasma coupling factor which we assume to be 3%. In this relation if we use the DT fuel solid density, $0.2\,\text{g/cm}^3$, the necessary laser energy is about $10^9\,J$ which is out of range in the present laser scale. However when the fuel density is increased to 100 times of the solid density, the laser energy decreases to $100\,kJ$. It seems to be a more preferable scale of laser in the present art of technology.

Using a laser large enough to ignite the fuel, the center of a fuel core reaches the fusion temperature and the produced α particle begins to heat up the outer part of the fuel. The burning rate of the fuel is determined by the ratio of the burning velocity to the exploding velocity of the fuel pellet. For a simple fuel pellet the burning ratio is given by

$$f_B = \frac{\rho R}{6 + \rho R}$$

Then the generated fusion energy is as follows:

$$E_{out} = \frac{4}{3}\,\pi \rho R^3 f_B \times 4.2 \times 10^{11}\ [J]$$

The pellet gain is

$$Q = \frac{E_{out}}{E_L} = \frac{913\rho R\varepsilon}{(6 + \rho R)}$$

To get the scientific breakeven condition, $Q = 1$, the condition $\rho R = 0.2\,\text{g/cm}^2$ is needed. We consider a laser fusion reactor which has a larger pellet gain to keep the self-supported operation. The input energy for the laser E_{in} is given by

$$\frac{E_{out}}{E_{in}} = \eta_L Q \eta_T (0.8M + 0.2)$$

where η_L is the laser efficiency, η_T is the thermal efficiency of the heat engine and M is the reactor blanket gain. In this case we use the D-T reaction and the blanket is provided by various fissionable materials of the $14.3\,MeV$ neutron.

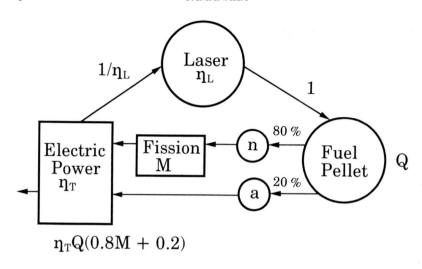

$$\eta_T Q(0.8M + 0.2)$$

Figure 2.2 Several concepts of fusion reactor.

As shown in Figure 2.2, if we set the laser efficiency $\eta_L = 5\%$, the pellet gain $Q = 50$, a pure fusion reactor is obtained $E_{out} = E_{in}$ at $M = 1$, where η_T is 40%. When we use the glass laser whose efficiency $\eta_L = 0.2\%$, the pellet gain $Q = 100$ and the blanket multiplication factor $M = 16$ are necessary for a fission-fusion hybrid reactor. The blanket material will be natural uranium for $M = 16$. In Figure 2.2, the energy circulating factor for the reactor operation is 100% in the internal circulation. This level corresponds the scientific breakeven condition.

3 FUSION DRIVER

3.1 Present Status of Drivers

The concepts and system designs of present drivers for inertial confinement fusion experiments are described. An overview of the current status of driver devices in research institutes in the world is given.

3.1.1 Lasers

(1) Glass laser
For glass lasers, Nd doped glass is used as the active medium. Since laser glass can provide a large stored energy as the population inversion, and a large scale laser glass with high homogeneity can be easily obtained, many high power glass laser systems have been constructed in several institutes for ICF experimental research (Krupke, 1974; Stokowski *et al.*, 1978; Yamanaka *et al.*, 1981a; Emmett *et al.*, 1984a; Hunt *et al.*, 1989). The principal high power glass laser systems are summarized in Table 3.1. The MOPA (Master Oscillator Power Amplifier) system is generally adopted where several hazards are solved. These are: parasitic oscillation in the amplifiers; the beam degradation due to the diffraction in the long optical path; and the amplification of the retroreflected laser pulse passing through the amplifiers. To avoid these kinds of problems, the laser amplifier chain should be provided an isolation between the amplifiers, the image relay to compensate the diffraction (Hunt *et al.*, 1978) and for rejection of the retro-pulse (Eidman *et al.*, 1972). According to these requirements, the laser amplifier chain usually consists of several functional components as described later. Here the GEKKO XII glass laser system of Osaka University is shown as an example of a large scale glass laser for fusion applications (Yamanaka *et al.*, 1986c).

1) GEKKO XII laser system
The 12 beam GEKKO XII glass laser system is shown in Figure 3.1 and the irradiating target system is also given in Figure 3.2. This experimental system consists of a laser amplifier chain, 12 beam switching mirror system and two target irradiation chambers. The laser amplifier chain has a large number of components such as the oscillator, the preamplifier chain, the beam dividing optics, the main amplifier chain, the laser alignment system, the laser control system and the laser energy monitoring system. The laser amplifier chain is set for the generation of high power laser pulses with good temporal and spatial profile. The beam switching system can swing the beams from one target chamber to another

9

Table 3.1 Large scale glass laser systems for ICF in the world.

Country	System	Laboratory	Beam number	Output power(TW)	Output energy(kJ)	Pulse width(ns)	Experimental Results
USA	NOVA	LLNL	10	100	120 (ω) 80 (2ω) 70 (3ω)	0.1~3.0	10^{13} neutrons
	OMEGA	LLE U.Rochester	24	15	3 (ω) 2 (ω)	0.1~1.0	10^{10} neutrons, ×200Liq.Dens.
	PHAROS-III	NRL	3	1.3	1.4 (ω) 0.8 (2ω)	0.1~1.0	
	CHROMA-I	KMS	2	0.6	1	0.1	
	MISHEN	Kurchatov	4	-	1	1.0	
USSR	DELFIN 2	Levedev	216	33	10	0.2~3.0	planning
	AURORA		20	-	50~500	0.03~10	
Japan	GEKKO-IV	ILE Osaka	4	2	1	0.1~1.0	10^{8} neutrons
	GEKKO-XII		2	7	2	0.1~1.0	2×10^{8} neutrons
	GEKKO-XII		12	55	30 (ω) 20 (2ω) 15 (3ω)	0.1~1.0	10^{13} neutrons, ×600Liq.Dens.
UK	VULCAN	Rutherford Appleton	12	3.6	5 (ω) 2 (2ω)	0.1~1.0	
France	PHEBUS	Limeil	2	20	20 (ω)	0.1~3.0	
	OCTAL		8	2	1	0.1~1.0	
	GRECO	Ecole Poly.Tec.	1	0.25	0.25	0.1~2.5	
China	SHEN GUANG	Shanghai	2	2	2	0.1~1	

Figure 3.1   Main amplifier chain of the GEKKO XII glass laser system.

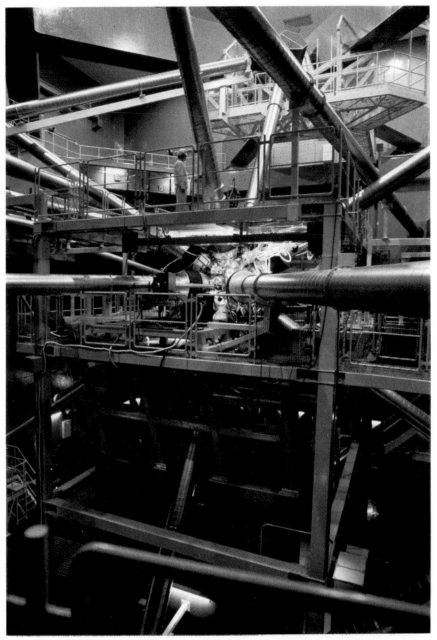

Figure 3.2   Target irradiation chamber of the GEKKO XII glass laser system.

in a few seconds. The target irradiation system is responsible for focusing the laser beam onto the two target chambers by green or blue laser light. This irradiation system consists of the tuning mirrors, the frequency conversion crystal, the random phase plate, the focusing lens, the beam energy monitor and the irradiation alignment system.

(i) Components of laser
As an example, details of the components of the GEKKO XII laser are described.

(a) Oscillator (OSC): In a glass laser system for ICF research, the oscillator is a key element for obtaining stable laser output pulses. Two mode locked Q-switch YLF (Yttrium Lithium Fluoride) oscillators with different pulse durations are used at the front end of GEKKO XII. The stability of the output pulse energy of the oscillator is better than ± 2% for 100 ps to 1 ns.

(b) Pulse Shape Modulator: For the purpose of obtaining an adequate laser pulse shape for target irradiation, a multimirror pulse stacking system is used in GEKKO XII to generate tailored laser pulses of various kinds (Kanabe et al., 1986).

(c) Amplifiers: In the laser chain, the rod type (RA) lasers are used in a smaller beam diameter (less than 50 mm), and the disk type (DA) lasers are used for 100 to 200 mm in beam diameter. The laser glass is Nd doped phosphate glass (Hoya LHG- 8; doping density $0.6 \sim 2\%$) and the transmission wave front distortion is better than $\lambda/6$.

(d) Optical Shutter (OS): To eliminate parasitic oscillations in the amplifier chain, optical shutters are installed in the necessary positions of the chain. These optical shutters consist of a Pockels cell and one pair of thin film multilayer coated polarizers, and a high speed, high voltage pulse generator which drives the Pockels cell.

(e) Spatial Filter (SF): The spatial filter in the laser chain plays three important roles by (1) selecting the spatial mode of the laser beam (Bettinger et al., 1976), (2) relaying the image of the input laser beam pattern to the next spatial filter (Hunt et al., 1978), (3) expanding the beam for the next stage of the larger beam diameter. For this purpose, the incident laser beam is focused with an input lens and the spatial mode is selected by placing a pinhole at the focal plane of the input lens. The output lens of longer focal length collimates the beam and expands the beam diameter to the desired size. These two lenses provide the image relay of the input beam pattern to the downstream.

(f) Faraday Rotator (FR): This component is a laser beam isolator for rejecting the retroreflected laser pulses in the amplifier chain (Eidman *et al.*, 1972). The Faraday rotator consists of a Faraday glass plate placed in a uniform magnetic field of a solenoid coil and one pair of thin film multilayer coated polarizers in the input and output side at a mutual angle of 45°. The Faraday glass rotates the polarization of the input laser beam by 45° and the retro-pulse is also rotated by the same angle so as to be rejected with the input polarizer.

(ii) Power Supply
A large scale capacitor bank is used for the energy storage which is fired by ignitron discharge tubes. In the GEKKO XII, the capacitor bank consists of about 2000 pieces of 25 kV, 44 μF capacitors and stored energy of about 23 MJ can be charged within 3 minutes.

(iii) Laser Alignment
In the GEKKO XII laser system, many laser beam measuring optics have been installed to accurately align the laser beam. The required accuracy of the laser beam centering is 1% of the beam diameter throughout the whole laser system. The automatic alignment system uses large numbers of micro-computers and motorized mirror gimbal mounts. The position of the laser beam is detected with semiconductor centering detectors or TV cameras placed in appropriate points of the laser chain. A closed loop control system is used for centering the beam in the amplifier. The laser alignment can be performed fully automatically to keep the excellent operation.

(iv) Irradiation Alignment
The laser beam alignment onto the target is a key issue for the ICF experiment. Since the frequency conversion crystals are located just before the focusing lens on the target chamber in the GEKKO XII, the final pointing accuracy of the incident laser beam to the target chamber should be kept better than 10 μ rad for maintaining high conversion efficiency. The irradiation alignment is achieved using a dummy target at the center of the target chamber which is made of a gold coated steel ball with high sphericity. The laser beam is focused on to this target where the reflected laser beam propagates back in exactly the same path as the incident beam. This alignment procedure is monitored with a detector located just behind the first turning mirror.

(v) Frequency Conversion

In ICF experiments, shorter laser wavelength light has several advantages in the plasma interactions (Yamanaka et al., 1972; Nuckolls et al., 1972). Recently, large scale glass laser systems for ICF have frequency conversion systems using nonlinear optical crystals (KDP: Potassium Dihydrogen Phosphate: $KH_2 PO_4$) for the second, third and fourth harmonic conversion (Craxton, 1981). The KDP crystal has several merits in the availability of a large size, high homogeneity, low absorption coefficient and high damage threshold single crystal. In the GEKKO XII laser system, second (2ω) and third (3ω) harmonic generation system are installed to the first and second target chamber, respectively. The conversion efficiency is up to 80% for 2ω and 60% for 3ω.

(vi) Uniform Irradiation Schemes

In ILE, the random phase plate is invented to suppress the irradiation nonuniformity of the coherent laser beam (Kato et al., 1984). Several alternative methods such as induced spatial incoherence using echelons with a broadband oscillator (Lehmberg et al., 1987) and multi-lens irradiation concept (Den et al., 1986) are investigated.

(vii) Laser Energy Monitoring

The energy monitoring is important in the large scale laser system. The output energy is measured in several points of the laser chain avoiding the distortion of the laser beam. For this purpose, measurement of the transmitted beam through the mirror at the bending point of the laser amplifier chain is adopted. The residual reflection of the polarizer in the optical shutter (OS) or the Faraday rotator is used for the additional energy monitor in the chain. The higher harmonic light energy is also measured through the final pointing mirror using the surface reflection of the KDP crystal cell. The measured energy data is collected with a data acquisition unit (DAU) and transferred to the laser control room. The final balance of the second harmonic energy in 12 beams on the target is better than $\pm$ 3% in GEKKO XII.

2) NOVA

The NOVA laser system, which was constructed in the Lawrence Livermore National Laboratory in the U.S.A., has 10 beams of 74 cm final beam diameter whose specification is 120 kJ (ω), 80 kJ (2ω) and 70 kJ (3ω) of laser light (Speck et al., 1986). The diameter of the final amplifier is 46 cm and split disks of laser glass are used for suppressing parasitic oscillations and for reduction of construction costs. However, this laser system was fatally damaged by the platinum inclusion in the laser glass and the output energy was reduced to $20 \sim 30\,kJ$ at 3ω for 3 years. Very recently, the laser glass has been replaced with a newly developed platinum-free laser glass (Campbell, 1986).

3) OMEGA

The OMEGA laser system of the Laboratory for Laser Energetics at the University of Rochester has 24 beams with the output energy of $3\,kJ$ at 3ω wavelength (Skupsky *et al.*, 1987). This laser system is modest but is suitable for the direct irradiation experiment. Recently, several attempts have been made to increase the irradiation uniformity.

(2) Excimer Laser

Since a short wavelength laser has several advantages in laser plasma interaction, a great deal of effort has been devoted to develop large scale excimer laser systems as an ICF driver (Suda *et al.*, 1984; Hunter *et al.*, 1986). The krypton fluoride (KrF) excimer laser of wavelength $\lambda = 248\,nm$ is promising for an ICF driver. Since the upper level of this laser is an excimer level which exists only in the excited state molecule and the lower laser level is a dissociative ground state, the bound-free transition provides efficient laser action. In an electron beam pumped excimer laser system, the energy deposited into the laser gas is converted to the population of the upper laser level with an efficiency of $10 \sim 12\%$ for a wide range of pumping power densities $(0.1 \sim 1\,MW/cm^{3})$. The demerit of this laser is the short lifetime of the upper level due to the large emission cross-section which demands the high pumping power to compensate the lack of energy storage. The laser pulse is determined by the pumping pulse duration. The nanosecond laser pulse is a key for ICF, so the light pulse compression is essentially important for the KrF laser. The size of the active medium of an amplifier is limited by the occurrence of amplified spontaneous emission (ASE). The scaling law of the e-beam pumped KrF laser given by Ueda *et al.* (1988) shows that the laser output energy, the volume of the active medium and the duration of the pumping pulse should be increased instead of decreasing the pumping power density.

In Table 3.2, the basic properties of the high power KrF excimer lasers are summarized.

1) AURORA

This laser system consists of a large aperture $(1 \times 1 \times 2\,m)$ electron beam pumped amplifier as the final power amplifier and the angular multiplexing system for the pulse compression. The final amplifier is pumped with four electron guns with an acceleration voltage of $675\,kV$ and a beam

Table 3.2 Basic properties of high power KrF excimer lasers.

Country	System	Laboratory	Laser energy (J)	Pulse width (ns)	Active volume (m³)	E-beam energy (keV)	E-beam current (kA)	Pulse-forming line
USA	AURORA	LANL	10.5 k	500	1×1×2	675	1000	4×SL 2.7Ω
UK	SPRITE	Rutherford Appleton	160	60	0.25Φ×1	450	360	4×SL 5Ω
Canada		Alberta Univ.	30	60	0.1×0.1×0.5	300	120	2×BL 5Ω
Japan	—	ISSP, Tokyo Univ.	160	70	0.23×0.23×0.8	400	270	2×BL 3Ω
	ASHURA	Electro-Technical Lab.	230	100	0.2×0.2×0.6	450	200	2×SL 4.6Ω
	—	ILS, Univ. Electro-Communications	83	120	0.11×0.11×0.7	350	70	2×BL 10Ω

current of 250 kA. This amplifier can generate 10 kJ of the output energy when an unstable resonator is used. The angular multiplexing has been provided by the use of a multimirror encoder-decoder system as shown in Figure 3.3. The Aurora laser has produced an optically multiplexed train of 96 beams 5 ns pulses containing approximately 2.5 kJ from the final amplifier and transported 0.78 kJ in 48 beams (equivalent to 1.56 kJ in 96 beams) to the target chamber (Rosocha et al., 1989).

2) ASHURA

This laser system has been constructed in the Electro-technical Laboratory of Japan for plasma experiments. A laser output energy of 660 J has been obtained with 4 beam angular multiplexing of 22 ns pulses. The laser output efficiency of the final amplifier reaches 2% for multiplexed pulse output. This system is operated at the energy fluence level of 1.5 J/cm^2 and the target irradiation experiment has been started (Owadano et al., 1989).

3) SPRITE

SPRITE has a cylindrical pumping system with 4 electron guns surrounding the laser gas for uniform excitation (Shaw et al., 1982). The beam diameter is 25 cm and 160 J output can be obtained for a 60 ns pulse. Although this laser system was constructed in the early stage of the development of this type of laser, the system is well designed and reliable operation can be obtained. Recently, a pulse compression system using stimulated Raman scattering of hydrogen gas has been introduced.

(3) Iodine Laser

The wavelength of the iodine laser is 1.315 μm and the frequency conversion technique using the KDP crystal can be achieved. Since this laser medium is in the gas phase, there are some advantages such as the rapid cooling of the lasing gas and a small nonlinear refractive index. A high power iodine laser system has been developed at the Max-Planck Institute for Quantum Optics in Germany (Witte et al., 1981). In 1978, the ASTERIX III laser system with 300 J output energy at 300 ps pulse was completed. Recently, an additional amplifier has been installed as ASTERIX IV and 5 TW/2 kJ output in 0.2–3 ns pulse duration will be delivered.

(4) CO$_2$ Lasers

The CO$_2$ laser has many advantages in generating a high energy laser beam as follows, (a) a large volume of active medium can be easily excited by the electron beam sustained high pressure discharge, (b) a direct

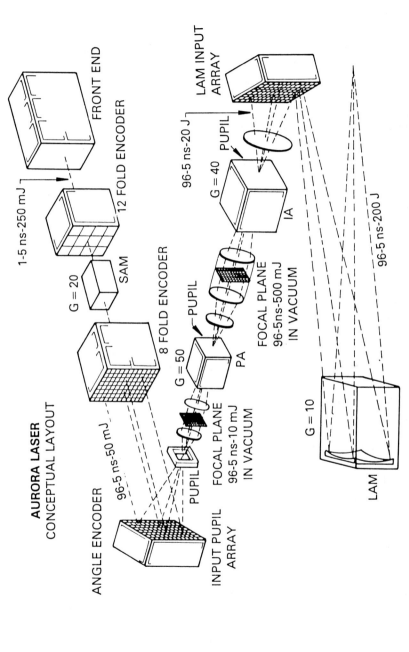

Figure 3.3 Layout of the AURORA excimer laser system. (*Courtesy of Los Alamos Scientific Laboratory.*)

excitation of the laser gas by the discharge provides high efficiency, (c) high speed laser gas flow permits a high repetition rate operation. Furthermore, the intrinsic efficiency due to the quantum level is sufficiently high ($\sim$ 40%). According to these superior characteristics, the CO_2 laser used to be thought of as the most probable laser for a fusion driver (Yamanaka *et al.*, 1981b; Carlson *et al.*, 1981). However, it was found that the wavelength of the CO_2 laser ($\sim$ 10 μm) is too long for the efficient implosion of the fusion target due to the hot electron generation in plasma.

A new concept of MICF (Magnetically insulated inertial confinement fusion) was proposed by Hasegawa *et al.* (1988) and the experiment has been performed at ILE, Osaka. The CO_2 laser will be useful in this concept.

3.1.2 Particle Beams

As discussed in the previous section, the laser drivers have advantages for generating short pulses (order of ns) and high peak power (up to 100 TW) with good focusability for small fusion targets (less than 1 mm). However, the laser devices have several disadvantages such as high construction cost, low efficiency and difficulties for obtaining the larger output energy. On the other hand, the pulsed power technology can easily generate a large amount of energy with high efficiency and low cost. It may provide a light ion beam for an ICF driver if the ion beam could be bright enough and transported to the fusion target efficiently.

In this section, as laser alternatives, light ion beams (LIB) are briefly discussed.

Since the light ion beam accelerator has the advantage of larger energy deposition rate onto the fusion target as compared with a relativistic electron beam (REB), attempts have been made for developing large scale LIB systems in many countries as shown in Table 3.3. For the ion beam diode to produce a bright source is essentially important. Various concepts are proposed (Goldstein, 1978; Johnson *et al.*, 1979). The beam transport to the target is also a crucial issue.

1) REIDEN IV
REIDEN IV is a light ion beam system with an output of 1 MV, 1 MA at 60 ns pulse and can generate a proton beam of 0.2 TW (Imasaki *et al.*, 1984). The output voltage is increased by use of an impedance transformer line (1 Ω to 9 Ω) and the pulse compression in the diode to 6.4 MV. An additional induction linac system has been connected to this system for generating 2 $\sim$ 10 MV output.

Table 3.3 Specifications of large scale LIB systems for ICF application.

Country	System	Laboratory	Output Parameter	Ion Beam Output	Diode	Remarks
USA	PROTO I	Sandia (1975)	2.0 MV, 0.5 MA, 24 ns	0.5TW	MID	Foil Cone Target Classical Interaction
	PROTO II	Sandia (1876)	1.5 MV, 6 MA, 24 ns	—	—	—
	PBFA I	Sandia (1980)	2.4 MV, 17 MA, 24 ns	—	—	—
	PBFA II	Sandia (1985)	10 MV, 3 MA, 40 ns	—	—	Disc Target, Classical Interaction and Hydrodynamic Properties
	GAMBLE II A	NRL (1974)	1.0 MV, 1.3 MA, 80 ns	0.4TW	Pinch-Reflex	
	HERMES III	Sandia (1988)	20 MV, 800 kA	—	—	—
USSR	ANGARA V-M	Kuruchatov (1979)	2.0 MV, 4 MA, 80 ns	—	—	8module
	ANGARA V	Kuruchatov	(2.0 MV, 25 MA, 80 ns)	—	—	plan
FRG	KALIF	KfK(1981)	1.9 MV, 0.8 MA, 50 ns	0.5TW	Pinch-Reflex	
France	SIDNIX	Valduc	1.3 MV, 0.7 MA, 80 ns	—	—	
Japan	REIDEN IV	Osaka (1979)	1.0 MV, 1.0 MA, 60 ns	0.2TW	Pinch-Reflex	
	REIDEN SHVS		4 MV, 40 kA, 100 ns	0.1TW	MID	Thin Film Target Classical Hydrodynamics
	ECHIGO	Osaka (1986) Nagaoka (1980)	1.0 MV, 0.2 MA, 50 ns	0.01TW	MID	—

2) PBFA II and HERMES III

PBFA II of the Sandia National Laboratories in the U.S.A. is the world's biggest pulsed power machine with the design value of $100 \sim 150$ TW, 3.7 MJ, $4 \sim 30$ MV output (Turman *et al.*, 1985). This system is the first machine with an output energy exceeding 1 MJ and the final goal is the achievement of ignition of the ICF target. The present operation level is 140 kJ, few TW/cm^2 and the full power is expected to be 1 MJ and $10 \sim 100$ TW/cm^2. The HERMES III induction accelerator is designed for the technology development of the laboratory microfusion facility (LMF) (Ramirez *et al.*, 1987). This system is currently operated as an electron accelerator of 600 kJ output and the full operation condition will be 400 kJ Li + ion beam generation.

3.2 Advanced Driver Development

Requisite performances of ICF driver for power reactors are shown in Table 3.4. Brief descriptions of these requirements are given as follows.

3.2.1 Conditions for reactor driver

(1) Output energy $(1 \sim 10$ MJ)
The theoretical consideration based on the experimental results indicates that $1 \sim 10$ MJ of the driver output energy is required for obtaining a high gain implosion. This output energy range is chosen as the goal of Laboratory Micro Fusion (LMF) program in the U.S.A. (Hunt *et al.*, 1987). This range of the driver output can be assumed as almost the same for different types of drivers.

Table 3.4 Requirements of drivers for ICF reactor.

Output energy	$1 \sim 10$ MJ
Efficiency	> 10 %
Pulse length	$1 \sim 10$ ns
Pulse shape	Adequate shape
Wavelength	$0.2 \sim 0.5$ μ m
Repetition rate	> 1 Hz

(2) Efficiency (> 10%)

In a power reactor system, the recirculation fraction of the energy flow should be less than 25% for the economical operation. When the thermal efficiency for electric power generation is assumed to be 40% and the pellet gain to be 200, the driver efficiency must exceed 10%.

(3) Pulse length (1 ~ 10 ns) and pulse shape

For efficient hydrodynamic implosions, an adequate power density on the target exceeding $100 \, TW/cm^2$ is required. The driver pulse must have a carefully designed pulse shape for obtaining ignition temperature and necessary density profiles in the implosion target.

(4) Wavelength (0.2 ~ 0.5 μm) or particle energy

The driver beam must provide the energy deposition on the target without the fuel preheat for an efficient implosion. The wavelength or the particle energy must be chosen so as to eliminate the generation of hot electrons or penetrating X-rays.

(5) Repetition rate (> 1 Hz)

When the output energy of the driver system is fixed, the output level of the electric power of the reactor system is proportional to the pulse repetition rate. This rate should be determined by taking into account the desired power level of the reactor, the resistance of the reactor chamber for the fusion blast and the recovery time of the driver.

3.2.2 Advanced Laser

(1) Solid state laser

The Nd glass laser is the best developed laser for fusion experiments. However, several problems still remain for realizing the reactor driver. The first is the heat removal from the solid state active medium under the high repetition operation condition in a large scale laser system (Emmett et al., 1984b). The cooling scheme of the laser glass is an important issue of the glass laser development. The second is the achievement of high efficiency sufficient for the reactor driver. The flash lamp pumping provides too much heat deposition onto the active medium in high repetition rate operation. A new pumping scheme using the semiconductor laser has been investigated intensively (Scifres et al., 1982). Further efforts are being made for developing new crystalline laser material which is suitable for cooling.

Several attempts for developing advanced solid state lasers have been planned and tested. In Japan, ILE Osaka is developing the 100 kJ output

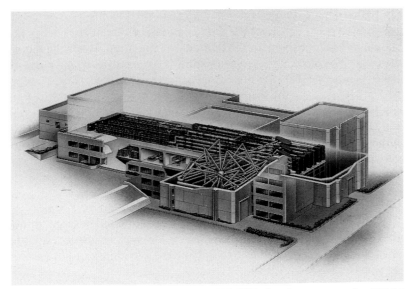

Figure 3.4 Up-grade system of the GEKKO XII glass laser system for 100 kJ at 3 ω.

laser system at 3ω wavelength for the breakeven or ignition experiment. This laser system can be constructed by increasing the beam numbers and additional booster amplifiers to the present GEKKO XII laser system. An artist's drawing of this laser system is shown in Figure 3.4. And a high repetition slab laser system has been constructed for the basic research of the solid laser development (Kanabe *et al.*, 1989). The diode pumped solid state lasers are also investigated for the development of the advanced laser.

In the U.S.A., wide range researches for developing high power and high average power laser systems have been made intensively in Lawrence Livermore Laboratory. The ATHENA laser program is proposed to prove the feasibility of the solid state laser for the reactor driver. ATHENA is designed to have a laser output energy of 10 MJ at short wavelength of 10 ns pulse duration (Hunt *et al.*, 1987). The current design is based on the flash lamp pumping and the multipass amplification scheme in 100 beams. Another effort is the HAP (high average power) laser project which is using new laser materials (Gelinas *et al.*, 1987). In Rochester, an upgraded OMEGA laser system with 60 beams and 30 kJ output energy has been designed (Kelly *et al.*, 1989).

(2) Gaseous laser

The KrF laser system has the intrinsic advantage of a short wavelength. However, since this laser is suitable for energy extraction in long pulse duration, a complicated angular multiplexing scheme must be used. The effectiveness of this scheme is not proved at the present time. Furthermore, the power amplifier is pumped with the electron beam through the window foil. Several difficulties may exist in the angular multiplexing of a large number beams of large aperture, the cooling of the laser gas without the turbulence, and the durability of the window foil for the repetitive operation in the reactor.

In the U.S.A., the AURORA laser facility is under construction and is currently being tested. The design work of a reactor driver is also made. In Europe, a new laser system "SUPERSPRITE" for 3.5 kJ output with 12 beams is designed for the next step. A plan of constructing 100 kJ laser system as the EURO-LASER program has been discussed in European institutes very recently.

(3) Free electron laser

The free electron laser (FEL) is based on the light amplification due to the resonant interaction between the electron beam and the magnetic field. Since the wavelength of FEL is intrinsically variable and the laser efficiency is in principle high, a high power FEL is expected to be developed for ICF applications.

The FEL system consists of an electron beam accelerator, a periodic magnetic field (Wiggler) and an optical resonator. The high intensity electron beam of relativistic energy is injected into the Wiggler field and the electron beam is bent by the periodic field so as to generate the radiation. The pulses of the radiation and the bunch of electrons transit synchronously through the Wiggler with a suitable timing. The light pulse is resonated in the optical cavity and the electron beam is recirculated from the Wiggler to an electron cooling device for efficient operation. The wavelength of the FEL can be varied by changing the electron beam energy and the strength or the pitch of the Wiggler field.

The first demonstration of the FEL laser at a wavelength of $3.4 \mu m$ was made in 1976 at Stanford University (Deacon et al., 1977). Recently, a FEL program started at the Institute for Laser Technology, Osaka for energy application. Large scale induction accelerators have been constructed using pulse power technology in several institutes for FEL applications (Cook et al., 1983; Imasaki et al., 1989). Figure 3.5 shows the induction linac FEL in the Institute of Laser Engineering, Osaka University.

Figure 3.5 Induction linac free electron laser in ILE Osaka University.
(*above*) Induction linac 9 MeV, 3 kA.
(*below*) Wiggler wavelength 6 cm, 30 pitches.

3.2.3 Heavy Ion Beam

The rapid development of ICF research using lasers and LIB stimulated the application of the heavy ion beam to the ICF driver (Faltens *et al.*, 1979; Fessenden *et al.*, 1987). This scheme largely depends on the accelerator development by which principally high repetition rate and high efficiency operation are possible. The present output of such a heavy ion beam accelerator is low and the cost is very high. An ambitious program has been performed for constructing the heavy ion RF linac and the storage ring system in GSI of West Germany (Muller, 1988). The installation is shown in Figure 3.6.

The induction linac is another promising accelerator of heavy ion

Figure 3.6   RF linac for heavy ion beam ICF. (*Courtesy of G.S.I.*).

because the output power is high and the cost of the hardwear is relatively low. In Lawrence Berkeley Laboratory, the output energy is designed to be 3.3 MJ, and the beam transport experiment and development of the induction linac are currently being conducted (Fessenden *et al.*, 1979).

4 LASER PLASMA INTERACTION

Laser light irradiating on a fusion target induces the evaporation of the target material. As the laser light flux is as high as 10^{14} W/cm^2, a high temperature plasma is produced which is an electrically almost neutral mixture of ions and electrons.

The electric field of the laser light makes a quivering motion of electrons which is thermalized by the collision with ions. This is the well-known inverse-Bremmstrahlung effect, often called classical absorption.

Increasing the laser intensity, the plasma temperature rises too high to make collisions between particles which leads to a low electric resistivity, that is, the weak absorption of the laser light. Actually the glass laser light of wavelength 1.06 μm begins to decrease the absorption beyond the laser intensity 10^{13} W/cm^2 owing to a decrease in collisions. Laser light impinging into a plasma with a density gradient reaches the cut off density region and then reflects back from there.

If the laser intensity increases to much higher than a certain threshold the absorption increases again. This is so-called anomalous absorption which is introduced by nonlinear effects of coupling between the laser light and the plasma waves. This phenomena, which causes the generation of hot electrons, was experimentally found by the author. (Yamanaka, 1972). In the 1970s the anomalous absorption became the main theme of laser plasma research. The parametric instability, Brilliouin scattering and Raman scattering due to the couplings between plasma waves and laser were intensely investigated. Obliquely incident laser light shows a strong resonant absorption when the laser light turns around at the cut-off point where the direction of the light electric field coincides with the direction of the density gradient to produce the plasma oscillation. Figure 4.1 indicates general features of the laser light interaction with an inhomogeneous density plasma.

4.1 Classical absorption

The collisional absorption is a fundamental mechanism of the light absorption in plasmas. It is well understood in various cases (Kruer, 1988).

4.1.1 Homogeneous plasma

The light wave induces the quivering motion of electrons which is moderated by collisions with the background ions. For Maxwellian distribution of electron velocities, the electron ion collision frequency ν_{ei}

29

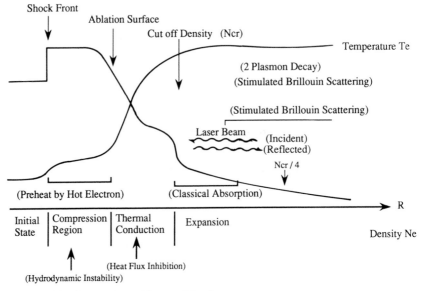

Figure 4.1 Interaction of laser with plasma.

is used to estimate the linearized force equation for the homogeneous electron fluid, (Johnston *et al.*, 1973)

$$\frac{\partial V_e}{\partial t} = -\frac{e}{m} E - \nu_{ei} V_e \qquad (4.1.1)$$

where V_e is the electron velocity and E is the electric field of the light wave. Since the form $E \cdot e^{-i\omega t}$ varies harmonically

$$Ve = \frac{-i_e E}{m(\omega + i\nu_{ei})} \qquad (4.1.2)$$

The electron current density is

$$J = -n_e e V_e = \frac{i\omega_{pe}^2}{4\pi(\omega + i\nu_{ei})} E,$$

where ω_{pe} is the electron plasma frequency. Then the plasma conductivity is complex,

$$\sigma = \frac{i\omega_{pe}^2}{4\pi(\omega + i\nu_{ei})}, \qquad \omega_{pe} = \left(\frac{4\pi n_e e^2}{m}\right)^{\frac{1}{2}} \qquad (4.1.3)$$

where $J = \sigma E$ is used. Maxwell equations for this plasma are

$$\nabla \times E = \frac{i\omega}{c} B \tag{4.1.4}$$

$$\nabla \times B = \frac{4\pi}{c} \sigma E - \frac{i\omega}{c} E = - \frac{i\omega}{c} \varepsilon E \tag{4.1.5}$$

where the dielectric constant of plasma is given by

$$\varepsilon = 1 - \frac{\omega_{pe}^2}{\omega(\omega + i\nu_{ei})} \tag{4.1.6}$$

The wave equation for E is obtained by (4.1.4) and (4.1.5),

$$\nabla^2 E - \nabla (\nabla \cdot E) + \frac{\omega^2}{c^2} \varepsilon E = 0 \tag{4.1.7}$$

For a spatially uniform plasma, we can get the dispersion relation for light wave, substituting $E(r) \sim e^{ik \cdot r}$ and ε to (4.1.7)

$$\omega^2 = k^2 c^2 + \omega_{pe}^2 \left(1 - \frac{i\nu_{ei}}{\omega} \right) \tag{4.1.8}$$

where $\nu_{ei}/\omega \ll 1$ is assumed. The light wave is damped, putting $\omega = \omega_r \sim i\nu_c/2$, where ν_c is the energy damping rate,

$$\omega_r = (\omega_{pe}^2 + k^2 c^2)^{\frac{1}{2}}$$

$$\nu_c = \frac{\omega_{pe}^2}{\omega_r^2} \nu_{ei} \tag{4.1.9}$$

The rate of energy loss of the light wave $\nu_c E^2/8\pi$ should be in balance with the rate of energy loss of quivering motion of electrons by the scattering due to ions, $\nu_{ei} n_0 m \nu_q^2/2$. Since $\nu_q = eE/m\omega_r$, the energy balance gives $\nu_c = \nu_{ei} \omega_{pe}^2/\omega_r^2$.

For the spatial behavior, let ω be real and k be complex. Substituting $k = k_r + ik_i/2$ to (4.1.8). We get at $k_i \ll k_r$

$$k_r = \frac{1}{c} \sqrt{\omega^2 - \omega_{pe}^2}$$

$$k_i = \frac{\omega_{pe}^2}{\omega^2} \frac{\nu_{ei}}{v_g} \tag{4.1.10}$$

where k_i is the rate of energy decay in space and v_g is the group velocity of the light wave. The energy damping length (k_i^{-1}) is given by v_g/ν_c.

4.1.2 Inhomogeneous plasma

(a) Normal incidence
When we assume a linear density profile $n_e = n_{cr} z/L$, where $n_{cr} = m\omega^2/4\pi e^2$, the wave equation is

$$\frac{d^2E}{dZ^2} + \frac{\omega^2}{c^2}\left(1 - \frac{z}{L}\right)E = 0 \qquad (4.1.11)$$

The variables are changed as

$$\xi = \left(\omega^2/c^2L\right)^{1/3}(z - L)$$

Then

$$\frac{d^2E}{d\xi^2} - \xi E = 0 \qquad (4.1.12)$$

This equation defines the well known Airy functions A_i and B_i. The general solution is

$$E(\xi) = \alpha A_i(\xi) + \beta B_i(\xi)$$

where α and β are constants determined by the boundary conditions. E should be a standing wave for $\xi < 0$ and decaying as $\xi \to \infty$ as $B_i(\xi)$ becomes infinite when ξ approaches ∞, β should be zero. The function of $A_i(\xi)$ is shown in Figure 4.2.

The wavelength as well as the amplitude of E increases as the reflection is approached. The constant α is determined by the incident wave at the interface of the plasma and the vacuum at $z = 0$, $\xi = -(\omega L/c)^{2/3}$. When the scale length of density is large, $\omega L/c \gg 1$, the asymptotic form is chosen

$$A_i(-\xi) = \frac{1}{\sqrt{\pi}\eta^{1/4}} \cos\left(\frac{2}{3}\eta^{1/3} - \frac{\pi}{4}\right)$$

As shown in Figure 4.2, the electric field has swelling at $\xi = 1$ and increases up to $3.6(\omega L/c)^{1/3}$ times of the amplitude of the incident wave. The light wave is collisionally damped in an inhomogeneous plasma. If we assume a linear density profile $n_e = n_{cr} z/L$ and a fixed value of collision frequency ν_{ei}, the dielectric constant is given by

$$\varepsilon = 1 - \frac{z}{L(1 + i\nu_{ei}/\omega)} \qquad (4.1.13)$$

The energy decay of the laser wave is calculated by the wave equation using an analytic solution with a linear density gradient and also from a WKB method.

Analytic Solution with Constant Density Gradient

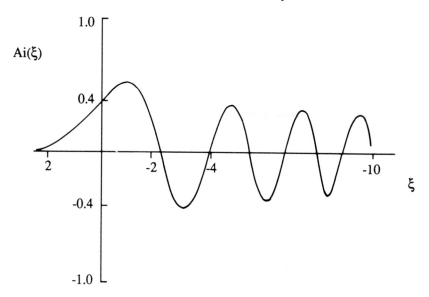

A plot of the Airy function $A(\xi)$

Figure 4.2   Airy function $A_i\ (\xi)$.

The absorption coefficient η_{ab} is given by

$$\eta_{ab} = 1 - \exp\left(-\frac{8}{3}k_0 L \frac{\nu_{ei}}{\omega_0}\right) \qquad (4.1.14)$$

where ω_0, k_0 is the laser frequency and the wave number respectively.

(b)   Oblique incidence

A plane light wave incident onto an inhomogeneous plasma with an angle θ is shown in Figure 4.3.

The electric vector of the light wave has two configurations of S-polarized and P-polarized which are vertical and parallel to the plane of incidence respectively. If the light wave is S-polarized $E = E_x x$, the wave equation becomes

$$\frac{\partial^2 E_x}{\partial y^2} + \frac{\partial^2 E_x}{\partial z^2} + \frac{\omega^2}{c^2}\varepsilon(z)E_x = 0 \qquad (4.1.15)$$

As the dielectric constant is a function of z alone, the wave number $k_y = (\omega/c)\sin\theta$ and $k_z = (\omega/c)\cos\theta$.

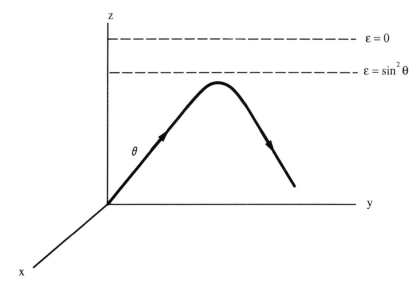

Figure 4.3   Oblique incidence of laser to inhomogeneous plasma.

$$E_x = E(z)\exp(i\,\omega y\,\sin\theta/c) \qquad (4.1.16)$$

Then

$$\frac{d^2E(z)}{dz^2} + \frac{\omega^2}{c^2}(\varepsilon(z) - \sin^2\theta)E(z) = 0$$

The reflection occurs at $\varepsilon(z) = \sin^2\theta$ and $\omega_{pe} = \omega\cos\theta$ as $\varepsilon = 1 - \omega_{pe}^2(z)/\omega^2$. An obliquely incident light wave reflects at a lower density than the critical density where $n_e = n_{cr}\cos^2\theta$. The critical density n_{cr} is given by $m\omega^2/4\pi e^2$. We can estimate the absorption coefficient of the laser light in the same way of the normal incidence case of a linear density distribution $n_e = n_{cr}z/L$ by the direct replacement $\varepsilon(z)$ to $\varepsilon(z) - \sin\theta$ and the reflect point $z = L$ to $z = L\cos^2\theta$.

$$\eta_{ab} = 1 - \exp\left(-\frac{38}{15}k_0L\frac{\nu_{ei}}{\omega_0}\cos^5\theta\right) \qquad (4.1.17)$$

To determine the absorption coefficient, the electron collision frequency must be estimated which depends strongly on the electron temperature T_e. The balance of the laser heating and the thermal conduction loss has a key to determine the electron temperature. Then we can put the thermal flux of the plasma as $f n_{cr} V_e T_e$, where f is a so called thermal flux limiting factor. The balance of the energy flow gives the electron temperature,

$$f n_{cr} V_e T_e = I_0 \eta_{ab}$$

From this relation we can estimate the dependence of the absorption coefficient upon the incident laser intensity I_0. As an example, Figure 4.4 shows the estimated absorption coefficient related to the laser intensity with the density scale length L = 100 μm and the laser wavelengths 1 μm, 0.5 μm and 0.35 μm which are well accorded with experimental results.

The classical absorption decreases with the increase of the laser intensity as well as the increase of the laser wavelength. In assuming an expanding plasma, a simple model yields the total absorption coefficient η_{ab} in the form (Max, 1982)

$$\eta_{ab} \propto \frac{f^{0.4} \left(z^{3/2} \tau \right)^{0.6}}{I_e^{0.4} \lambda_L^2}$$

where z is the ionization level, I_L is the laser intensity, λ_L is the laser wavelength, τ_L is the laser pulse duration and f is the thermal flux limiting factor.

4.2 Anomalous absorption

When the classical absorption decreases at the higher plasma temperature produced by the stronger laser intensity, the laser light proceeds to the

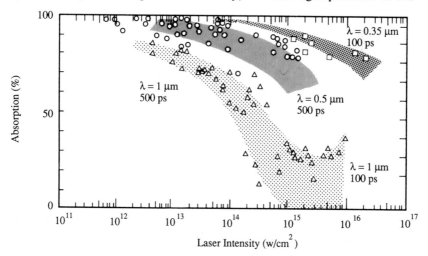

Figure 4.4 Absorption coefficient and its wavelength dependence.

cut off density region, where the conversion from the laser light to the plasma waves is introduced by the resonance absorption and parametric instability. The plasma waves are then damped by the collective motion of plasma particles. This kind of processes is called Landau damping. The mode conversions from the light wave to the plasma waves are specified into two categories: a linear process and a nonlinear process.

4.2.1 Resonance absorption

The resonance absorption is introduced by the obliquely incident P-polarized light wave when the density scale length is short in plasmas. In this case a component of the electric vector oscillates in the direction of the density gradient. As $E \cdot \nabla n_e \neq 0$, this oscillation generates fluctuations in plasma which can resonantly induce the electrostatic wave in plasma.

A plane electromagnetic wave incidents onto an inhomogeneous plasma with an angle θ as shown in Figure 4.3. This mode conversion from the light wave to the plasma wave finally heats up the plasmas due to the collective interaction between the plasma waves and the electrons. Since the energy is given selectively to suitable parts of the electrons, they obtain much higher temperature than that of the background electrons. They are called hot electrons and this absorption process is not acceptable to ICF, which intends compressing the fuel to an extremely high density by keeping the fuel cold during the laser irradiation.

Resonance absorption is formulated as follows. Maxwell equation to Ampere's law can be reduced with the equation of electron motion

$$m\frac{dv}{dt} = -eE,$$

$$\frac{d^2}{dt^2}v + \omega_{pe}^2(z)v = -\frac{e}{m}c(\nabla \times B)_z, \qquad (4.2.1)$$

where we assume the electron velocity v is in z direction and $j = -en_e v$ is used. In Eq. (4.2.1) the right part represents a driver term due to the tunneling of the laser field E_z and is non zero in the case of P-polarized incidence. Assuming $v \propto e^{-i\omega t} + c.c.$ and using the equation of electron motion, we can evaluate the absorbed laser power to be

$$I_{abs} = \int_{-\infty}^{\infty} j \cdot E \, dz,$$

$$= \frac{\omega}{8\pi}|\frac{c}{\omega}\nabla \times B|_{z=0}^2 \, Im\left(\int_{-\infty}^{\infty} \frac{1}{\epsilon(z)} \, dz\right) \qquad (4.2.2)$$

where $\epsilon(z) = 1 - \omega_{pe}^2(z)/\omega^2$ and $\epsilon = \sin^2\theta$ at the reflection point $z = L$ $\cos^2\theta$. By the use of the causality, the integration of Eq. (4.2.2) can be done in the complex z plane, resulting Im( ) $= \pi L$ with the scale length of the plasmas near the critical point, L. Finally, evaluating the driver term with the laser field in the vacuum (Gintzburg, 1970) yields the absorption coefficient as,

$$\eta_{res} = \frac{1}{2}\Phi^2(\tau) \tag{4.2.3}$$

In Eq. (4.2.3), $\Phi(\tau)$ is called Gintzburg curve and has a peak near $\tau = (k_0 L)^{\frac{1}{3}} \sin\theta = 1$, where k_0 and θ are the wave number of laser light in the vacuum and the incident angle, respectively. From Eq. (4.2.3), it is found that about 70% absorption is possible due to this type of absorption. Figure 4.5 is the Gintzburg curve for the resonant absorption.

When the dominant absorption by the above process is observed with

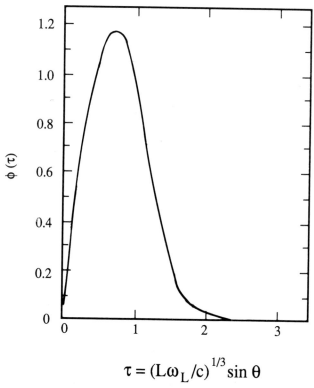

$$\tau = (L\omega_L/c)^{1/3}\sin\theta$$

Figure 4.5 Gintzburg curve for the resonance absorption.

the increase of $I\lambda_L^2$, the scale length of the expanding plasma, L, should be of the order of $k_0^{-1} = \lambda_0/2\pi$, which is, for example, $0.2\,\mu m$ for glass laser. It is very short compared to the scale length determined by the hydrodynamic motion. Actually, the resonance absorption plays an important role in the density profile modification due to the pondero-motive forces of the laser light coupled with the generated plasma wave. The ponderomotive force induces the density scale length near the critical surface to be short which enhances the resonance absorption (Estabrook, *et al.*, 1975; Takabe, *et al.*, 1982a).

Appearance of the hot electrons has been checked by means of particle simulations, and a scaling law of the hot electron temperature T_h has been given as follows (Forslund, *et al.*, 1977; Estabrook, *et al.*, 1978):

$$T_h \simeq 13 T_e^{1/4}(I_L\lambda_{L2})^{0.39}, (keV) \tag{4.2.4}$$

where T_e in keV and $I_L\lambda_L^2$ in 10^{15} W/cm^2 μm^2. To prevent the generation of hot electrons, the value of $I\lambda_{L2}$ must be kept roughly less than 10^{14} W/cm^2 μm^2.

4.2.2 Parametric instabilities

In laser produced plasmas, various parametric instabilities occur due to the couplings between waves. They are electromagnetic wave, electron plasma wave, and ion acoustic wave. When three of them satisfy the matching condition in energy and momentum, $\omega_0 = \omega_1 + \omega_2$ and $k_0 = k_1 + k_2$, then the wave energy of mode (ω_0, k_0) is converted to those of mode (ω_1, k_1) and mode (ω_2, k_2). According to the combination of three waves, the parametric instabilities are named as a) parametric decay instability, b) oscillating two stream instability, c) stimulated Brillouin scattering (SBS), d) stimulated Raman scattering (SRS), e) filamentational instability and self-focusing. Here only the stimulated scattering processes strongly related to the laser absorption are described.

When the ions are fluctuated with the density perturbations $\delta n_i (= \delta n_e)$, the electric field of the incident laser beam (E_0) produces a nonlinear current proportional to $\delta n_i E_0$, which then induces a scattered electromagnetic field E_1. If the ponderomotive force due to the beat wave produced by E_0 and E_1 satisfies the phase matching condition, a positive feed-back leads the growth of E_1 and δn_i to cause an instability.

When a plane-polarized electromagnetic field $E^{em} = E^{em}(x,t)\, i_y$ is propagating in the x-direction, we obtain from the Maxwell equation;

$$\left(\frac{\partial^2}{\partial t^2} - c^2\frac{\partial^2}{\partial x^2} + \omega_{pe}^2\right)E^{em} = -4\pi\frac{\partial}{\partial t}j_{NL,} \tag{4.2.5}$$

where the nonlinear current is given to be $j_{NL} = -e\delta n_e V^{em}$ and V^{em} is the quivering velocity of electrons by E^{em}. The density perturbation $\delta n_e (= \delta n_i)$ due to an ion acoustic wave is set by the following equation.

$$\left(\frac{\partial^2}{\partial t^2} + 2\Gamma_L \frac{\partial}{\partial t} - C_s^2 \frac{\partial^2}{\partial x^2}\right)\frac{\delta n}{n_0} = \frac{e}{m_i c^2}\frac{\partial}{\partial x} < v \times B >_x^{slow}, \quad (4.2.6)$$

where Γ_L and C_s are the damping constant and the sound velocity of the ion wave respectively and n_0 is the background plasma density. In Eq. (4.2.6), the ponderomotive force given in the right part shows a slowly varing component which can couple the ion wave oscillation. Assuming E^{em} consists of incident and reflected wave components $E_0 e^{-i\omega_0 t + ik_0 x} + c.c.$ and $E_1 e^{-i\omega_1 - ik_1 x} + c.c.$, respectively, and neglecting the second derivatives of the amplitudes, the following coupled equations are given;

$$\left(\frac{\partial}{\partial t} + c\frac{\partial}{\partial x}\right)E_0 = -i\alpha \frac{\delta n}{n_0} E_1,$$

$$\left(\frac{\delta}{\partial t} - c\frac{\partial}{\partial x}\right)E_1 = -i\alpha \frac{\delta n}{n_0} E_0, \quad (4.2.7)$$

$$\left(\frac{\partial}{\partial t} + \Gamma_L + C_s \frac{\partial}{\partial x}\right)\frac{\delta n}{n_0} = -i\omega_s \frac{E_1^* E_0}{4\pi n_{cr} T_e},$$

where $\omega_s = kC_s$, $\alpha = \omega_{pe}^2/2\omega_0$, and the relation $k_0 = k/2$ has been used. If a stationary state and strong damping of the ion wave are assumed Eq. (4.2.7) can be integrated to have a solution (Randal et al., 1979);

$$R(1 - R) = \mu\exp[Q(1 - R)],$$

where R is the reflectivity, and $Q = \alpha\omega_s |E_0(0)|^2 L/(4\pi n_{cr} T_e c_s)$ is called a quality factor; $E_0(0)$ is the amplitude of the incident wave at $x = 0$, L is the length of the system, and μ is the noise level of the reflected wave. In a case when $\mu = 10^{-2}$ and $Q = 5$, we obtain $R \simeq 50\%$.

The stimulated scattering process plays an important role in plasmas with a longer scale length. The incident laser is reflected back but can not reach the higher density region for depositing energy to produce hydrodynamic pressure.

The two plasmon decay instability and stimulated Raman scattering have also been studied extensively, because they produce super hot electrons. The former is induced by the decay of the incident laser into two plasma waves at the quarter critical density. In warm plasmas, the plasma waves with relatively long wavelength are predominantly excited.

Since the phase velocity of such waves is very large, super hot electrons are effectively produced.

The analytic treatment of SRS is similar to SBS where instead of the ion wave the electron plasma wave couples with the laser beam. In this case, the scattered electromagnetic wave can propagate in the backward and forward directions. The threshold intensity for SRS in forward is generally much higher than that for SRS in backward. However, SRS in forward produces the plasma wave with higher phase velocity, consequently generating a super hot electron with several MeV energy.

4.3. Energy transport for implosion

The energy transportation from the laser absorbed region to the ablation region is one of the most interesting topics for the implosion process. There are two kinds of transport owing to electrons and radiations.

4.3.1. Electron energy transport

Electrons heated by laser light are transported into the region with lower temperature and higher density, while the low temperature electrons come back, forming the heat conduction regions. Basically, the heat flux due to the electrons is described with a diffusion type approximation. However, the plasma is so abruptly heated up by the laser in the gradient length of temperature which is of the order of the electron mean free path. Then we need a new model for the heat transportation instead of the usual diffusion scheme.

As the transport phenomenon belongs to the global hydrodynamics, the flux limited model has been used in the fluid simulations for obtaining a good agreement with experimental results. In this model, the heat flux is;

$$q_e = \frac{q_{SH} \cdot q_l}{|q_{SH}| + q_l} \qquad (4.3.1)$$

where $q_{SH} = - K_e \nabla T_e$ and q_l is the limited flux defined by $q_l = f n_e v_e T_e$. The factor f is called flux limitation factor and is typically equal to $0.1 \sim 0.03$.

Heat conduction described by the random walk theory has the diffusivity of a form; $\chi = \lambda_e v_e /3$, where λ_e is the electron mean free path. The Spitzer's heat flux (Spitzer and Härm, 1953) is roughly given as;

$$q_{SH} \simeq \frac{\lambda_e v_e}{2} n_e \nabla T_e, \qquad (4.3.3)$$

in a uniform density plasma. Therefore, defining the temperature gradient scale length $L_T \equiv T_e / |\nabla T_e|$, the heat flux q_{SH} is written as;

$$q_{SH} \simeq \frac{\lambda_e}{2L_T} n_e v_e T_e . \tag{4.3.3}$$

From Eq. (4.3.3), it is easily understood that the heat flux for $L_T/\lambda_e < 1/(2f) \approx 0.1$ can not be described by the Spitzer's formulation.

Physical models of the flux limiter f have been attributed to various anomalous effects; ion wave turbulence (Manheimer, 1977) which reduces the electron mean free path λ_e, electrostatic field generation due to a hot electron driven return current (Takabe et al., 1982a), magnetic field generation, and so on. Although these models seem interesting, the study of transport has now focused on a more accurate treatment of the Fokker-Planck equation in a steep temperature gradient.

Several investigations on electron heat transport have been performed by solving the Fokker-Planck equation itself. Bell et al. (1981) have demonstrated that the heat flux is limited by about $0.1\ n_e v_e T_e$ and given a property of double valued functions to L_T/λ_e as shown in Figure 4.6.

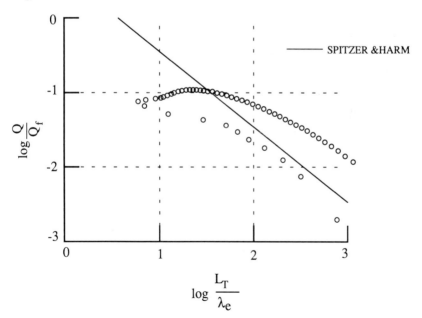

Figure 4.6 Electron heat flow Q to inverse temperature gradient where $Q_f = n_e v_e T_e$, L_T is temperature gradient scale length and λ_e is electron mean free path.

4.3.2. Radiation transport

In highly ionized plasmas, the energy transport by X-ray radiation appears to be very important compared to the electron energy transport. Radiation processes in medium or high Z plasmas essentially depend upon the atomic state of the plasmas. To study the atomic state of partially ionized plasmas it is required to deal with not only the radiation transport but also fluid dynamics including the equation of state, ionization level, and so on. The radiation transport has been studied extensively in astrophysics. (Zel'dovich *et al.*, 1966; Mihalas *et al.*, 1984).

Radiation transport is described by the equation;

$$\frac{1}{c}\frac{dI_\nu}{dt} = \varepsilon_\nu - k_\nu I_\nu \qquad (4.3.4)$$

where I_ν is the radiation intensity with the energy $h\nu$, ε_ν the emission rate, and K_ν the absorption coefficient. In assuming local thermodynamic equilibrium, the right part is given to be $K_\nu' (B_\nu - I_\nu)$, where $k_\nu' = k_\nu (1 - e^{-h\nu/kT})$ is the total absorption coefficient including induced emission and $B_\nu = 2h/c^2 \cdot \nu^3 /(e^{h\nu/kT} - 1)$ is a Planckian distribution. The absorption coefficient K_ν is due to free-free, bound-free, and bound-bound transition processes of the electronic state, and requires the detailed information of the electronic state in the plasmas and atoms (More, 1981). In solving radiation process in a fluid code, the averaged atom model or opacity tables are commonly used.

Equation (4.3.4) is solved numerically with the flux-limited multi-group methods (e.g., Wilson, 1972) or radiation temperature model (Zeldovich *et al.*, 1966). Since one can in general neglect the Compton scattering process in the laser plasmas, numerical treatment of Eq. (4.3.4) is relatively easier than that of the Fokker-Planck equation for electrons. In addition, as long as the condition $2\pi\nu \gg \omega_{pe}$ is satisfied, it is not necessary to consider the deflection effect and $d/dt = \partial/\partial t + v_g \partial/\partial r$ with $v_g = c$ is satisfied. From this point, the kinetics of radiation is much easier to treat than that of electrons affected by the self-generated fields.

5 IMPLOSION AND IGNITION

A scheme of target implosion driven by the laser irradiation is schematically shown in Fig. 5.1.

This is for the case of a shell fuel target and the time evolution of the implosion dynamics can be divided into four phases

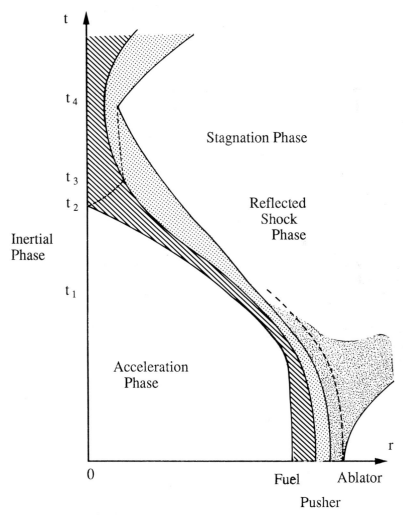

Figure 5.1 Schematic diagram of ablative driven implosion of a shell fuel target.

(1) Acceleration Phase ($0 < t < t_1$),
(2) Inertial Phase ($t_1 < t < t_2$),
(3) Reflected Shock Phase ($t_2 < t < t_3$),
(4) Stagnation Phase ($t_3 < t < t_4$).

In the acceleration phase (1), the laser is absorbed to ablate the outer region of the target material and the rocket repulsion drives the pusher and fuel to accelerate inward. In the inertial phase (2), the fuel and pusher freely fall toward the target center at an accelerated velocity V_A, the value of which is required to be $2–3 \times 10^7$ cm/sec (Bodner, 1981). In this phase, the fuel and pusher tend to expand due to their thermal pressure; therefore less preheating of these regions is necessary for the ideal ablative implosion. In the reflected shock phase (3), a void closure of shell triggers a shock wave which propagates outward, heating the fuel by converting directed kinetic energy to thermal energy. At the time when the shock wave arrives at the contact surface, the so-called stagnation phase (4) begins. If at $t = t_3$ the dynamical pressure (ρu^2) of the pusher is larger than the static pressure (P) of the fuel, such discontinuity of the pressures leads to the generation of transmitted and reflected shock waves. Then, the fuel is further compressed and gains energy from the pusher through adiabatic compression in the stagnation phase, if no mixing occurs.

5.1 Ablation process of pellet

When one considers the ablation of a low-Z material, the electron energy transport plays an important role. Ablated plasmas expand into the vacuum with the velocity of sound $C_s = (P/\rho)^{1/2}$. Then, the so-called deflagration form (Bobin, 1971; Takabe et al., 1978) is introduced in the heated region. As the ablation region has almost constant pressure, which is almost equal to the ablation pressure P_A, we can obtain an approximated energy conservation relation;

$$C_s \tau_L P_A \simeq E_{abs}, \qquad (5.1.1)$$

where E_{abs} is the absorbed laser energy, τ_L the pulse width of the laser and $C_s \tau_L$ is regarded as an effective width of the ablation region. In Eq. (5.1.1), we neglected the energy transferred to the accelerating shell. We obtain the ablation pressure in the form;

$$P_A \approx I_{abs}/C_s. \qquad (5.1.2)$$

It is interesting to note that the photon pressure without any absorption is $P_p = 2I_L/c$ and this pressure is C_s/c times smaller than the pressure generated in the ablation process. It is clear from Eq. (5.1.2) that for increasing the ablation pressure, keeping the sound speed as low as possible is preferable. The generation of the fast ions should be avoided and the usage of the drivers depositing the energies in the higher density region is requested.

If one assumes that the absorbed laser energy is carried by the electron heat flux $fn_e v_e T_e$ we have a relation $fn_e v_e T_e \simeq I_{abs}$, which then provides,

$$C_s \propto \left[I_{abs}/(n_{cr} f) \right]^{1/3} \tag{5.1.3}$$

Inserting Eq. (5.1.3) into Eq. (5.1.2), we get a scaling law of the ablation pressure;

$$P_A = P_0 f^{1/3} (I_{abs}/\lambda_L)^{2/3} \tag{5.1.4}$$

where P_0 is a constant and λ_L is the laser wavelength. In the model of stationary Chapman-Jougnet deflagration wave (Takabe et al., 1978; Ahlborn et al., 1982), the constant is shown to be

$$P_0 f^{1/3} = 12 \, \text{Mbar}, \tag{5.1.5}$$

where I_{abs} is in the order of $10^{14} \, \text{W/cm}^2$ and λ_L is in μm units in Eq. (5.1.4). In Figure 5.2, the ablation pressures experimentally obtained are plotted, where the solid lines show Eq. (5.1.4) with Eq. (5.1.5) for $\lambda_L = 0.53$ and $1.06 \, \mu m$. The mass ablation rate $\dot{m}$ is also calculated

$$\dot{m} = 1.5 \times 10^5 I_{abs}^{1/3} \lambda_L^{-4/3} \, (\text{g/cm}^2\text{sec}). \tag{5.1.6}$$

5.2. Rocket model for implosion

When the acceleration of the compressed region has the areal mass $M(t)$ and is accelerated to the velocity $V(t)$ at time t, the equation of motion is reduced to the well-known Rocket equation;

$$M(t) \frac{dV(t)}{dt} = P_A, \tag{5.1.7}$$

In obtaining Eq. (5.1.7), a plane geometry is assumed. Since $M(t) = M_0 - \int_0^t \dot{m} dt$, where M_0 is the initial mass, Eq. (5.1.7) is integrated with the assumption of constant $\dot{m}$ and P_A;

$$V(t) = 2C_s \ln\left(\frac{M_0}{M(t)}\right), \tag{5.1.8}$$

where the relation $P_A/\dot{m} = 2C_s$ is used.

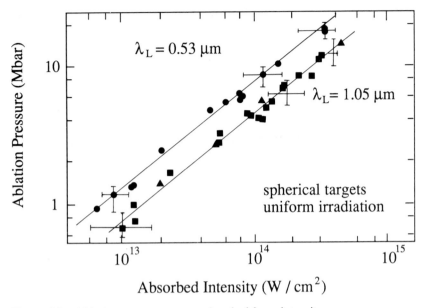

Figure 5.2   Ablation pressure versus absorbed laser intensity.

Hydrodynamic efficiency is defined by the ratio of the kinetic energy of the accelerated fuel divided by the absorbed laser energy. By relation $I_{abs} = 4\dot{m}C_s^2 \simeq 2P_A C_s$ and Eq. (5.1.8), the hydrodynamic efficiency η_H is given by

$$\eta_H = \frac{\frac{1}{2}M(t)V^2(t)}{I_{abs} \cdot t} = \frac{1}{2}\frac{\Phi(\ln\Phi)^2}{1-\Phi}, \qquad (5.1.9)$$

where $\Phi = M(t)/M_0$. It is found that η_H has its maximum at $\Phi \simeq 20 \sim 30\%$.

Now, we put $\Phi = 1/e(\approx 0.37)$ for obtaining peak value of η_H. Then, we get

$$V = 2C_s, \quad d = V^2 M_0/P_A \qquad (5.1.10)$$

where d is the travel distance. From Eq. (5.1.10), we find the following relation between P_A and V;

$$P_A = \rho_0 V^2\left(\frac{R_0}{d}\right)\left(\frac{\Delta R_0}{R_0}\right), \qquad (5.1.11)$$

where ρ_0 and ΔR_0 are the initial density and the thickness of a target shell respectively. Regarding R_0 as an initial target radius, $R_0/\Delta R_0$ is the intial aspect ratio of a target. As an example, assuming a target which consists of pure DT shell ($\rho_0 = 0.2 \, g/cm^2$), the required ablation pressure for obtaining $V = 3 \times 10^7 \, cm/sec$ is calculated from Eq. (5.1.11) to be 40 Mbar, where we set $R_0/d = 2$ and $R_0/\Delta R_0 = 10$. From Eq. (5.1.11), it is found that the larger aspect ratio ($R_0/\Delta R_0$), the less ablation pressure is necessary to accelerate the target. It is noted that from Eqs. (5.1.4) and (5.1.5) the laser intensity required for $P_A = 40 \, Mbar$ is $I_{abs} = 6 \times 10^{14} \cdot \lambda_{\mu m}$ (W/cm²). Then a condition for the classical absorption $I_L \lambda_{\mu m}^2 < 2 \times 10^{14} \, W/cm^2$ which prevents the generation of the hot electrons requires the wavelength $\lambda_L < 0.33 \, \mu m$. This is one of the reasons why shorter wavelength lasers have been prepared for implosion experiments.

5.3 Instability of ablation and stagnation

Instability in the implosion process is a crucial problem for the ICF. We have two cases of instability in the acceleration phase and in the deceleration phase. To prevent the growth of the instability is very important to pursue the implosion.

5.3.1 Ablation mode

It is important to analyze the Rayleigh-Taylor instability of the dynamics in order to answer whether the implosion scheme is realistic within the symmetry assumption. As shown in Figure 5.3, the steep density gradient is produced at $r = r_a$ against the inertial force working from the left to the right in the figure. Such a situation is unstable to the Rayleigh-Taylor instability, the growth rate of which is $\gamma = \sqrt{kg}$, where k is the wave number of a perturbation and g is the gravitation which is equal to the acceleration of the ablation front. However, recent studies (Takabe et al., 1983, 1985; Emery et al., 1982, 1986; McCrory et al., 1981) reveal the fact that the growth rate is reduced due to the ablation and the thermal conduction. An eigenvalue analysis in the self-consistent structure as shown in Figure 5.3, gave the growth rate in a tractable form;

$$\gamma = \alpha\sqrt{kg} - \beta k V_0, \qquad (5.1.12)$$

where $\alpha = 0.9$, $\beta = 3 \sim 4$ are constants and V_0 is the flow velocity across the ablation front, which is equal to $V_0 = \dot{m}/\rho_a$, where ρ_a is the maximum density at the ablation front.

In Fig. 5.4, the growth rate of Eq. (5.1.12) is given for CH-foil irradiated by 1.06 μm (solid lines) and 0.25 μm (dotted lines) laser lights.

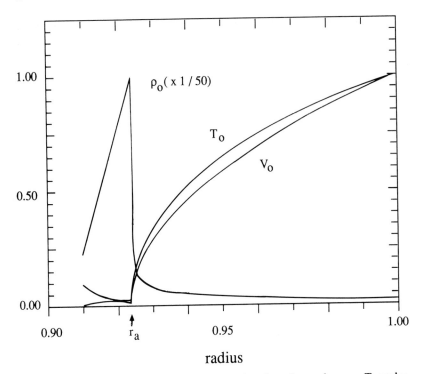

Figure 5.3 Target plasma profile in the acceleration phase where ρ_o, T_o and v_o are the density, temperature and flow velocity respectively. r_a shows the ablation surface.

Equation (5.1.12) suggests that if we use a driver and target materials which are able to enhance the ablative flow velocity V_0, we can reduce the growth of perturbations in longer wavelength and stabilize the instability with shorter wavelength.

Therefore, the usage of shorter wavelength laser and low solid density (low Z) material is a good choice against the Rayleigh-Taylor instability.

5.3.2 Stagnation mode

At $t = t_3$ in Fig. 5.1, we introduce a stagnation parameter defined by

$$X_s = \frac{\rho_{p,3}U_c^2}{P_{f,3}}, \tag{5.1.13}$$

where $\rho_{p,3}$, U_c, and $P_{f,3}$ are the density of the pusher, the velocity of the contact surface and the pressure of the fuel, respectively. If X_s is larger than 1, the fuel is further compressed almost adiabatically in the

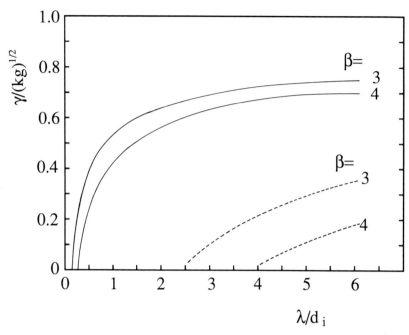

Figure 5.4 Growth rate of Rayleigh-Taylor instability at ablation front γ normalized by its classical value $\sqrt{kg}$ as a function of perturbation wave length λ divided by initial shell thickness d_i.

stagnation phase, until the time when the relation $U_c = \int^t \alpha_c dt$ is satisfied, where α_c is the deceleration of the contact surface due to the fuel pressure and this stagnation condition is to be;

$$C_{sp} = U_c, \qquad (5.1.14)$$

where C_{sp} is the sound speed of the pusher near the contact surface. Assuming the adiabatic process in the stagnation phase Eq. (5.1.14) provides the following relations;

$$T_{f,4} \simeq X_s T_{f,3}$$

$$\rho_{f,4} \simeq X_s^{2/3} \rho_{f,3} \qquad (5.1.15)$$

where $T_{f,4}$ and $\rho_{f,4}$ are the temperature and the density of the fuel at $t = t_4$, while $T_{f,3}$ and $\rho_{f,3}$ are these data at $t = t_3$. Here, it is useful to reduce Eq. (5.1.13) to the form;

$$X_s \simeq 3 \frac{\rho_{p,3}}{\rho_{f,3}} \qquad (5.1.16)$$

where the relation $3/2\,P_{f,3} = 1/2\,\rho_{f,3}U_c^2$ is used from the energy conservation. This is a good approximation if the internal energy before the shock generation can be negligible. As seen from Eq. (5.1.15), when the fuel is accelerated up to $Uc = 2 \times 10^7$ cm/sec, the fuel temperature, which is about 500 eV at t_3, will be 5 keV by stagnation heating, provided the pusher density is kept sufficiently high to satisfy $\rho_{p,3}/\rho_{f,3} \approx 3$. Then, the fuel density at the maximum compression is about 5 times the density at t_3.

In a stagnation dominant implosion mode, the fuel with lower density decelerates the pusher of higher density. The fuel-pusher contact surface tends to be unstable due to the Rayleigh-Taylor instability. This instability has been studied by Hattori *et al.* (1986), who solved an eigenvalue problem for linear perturbation from a self-similar solution which used the stagnation dynamics in a spherical geometry. They obtained a formula describing the temporal evolution of the perturbation amplitude $\xi(t)$ as;

$$\xi(t) \propto R_c(t)\exp\left\{\int^t [\alpha_A n(t)A(t)]^{1/2} dt\right\}, \qquad (5.1.17)$$

where $R_c(t)$ is the radius of the contact surface, α_A the Atwood number $n = l/R_c(t)$ the effective wave number of perturbation, and A the acceleration of the contact surface at t. For the case of $X_s \approx 16$, the temporal growths of the perturbation amplitudes for $\ell = 10, 20,$ and 40 are plotted in Figure 5.5. The ordinate is the value of $\xi(t)/R_c(t)$, while the abscissa is the normalized time taking $t = 0$ at the maximum compression. As shown in this figure, an explosive growth of the instability is seen in the stagnation phase, possibly inducing fuel-pusher mixing.

In order to avoid the instability accompanied by the stagnation dynamics, the implosion scheme characterized by $X_s < 1$ has been tested. For this purpose, it is required to ablate away the pusher until $t = t_3$ for decreasing $\rho_{p,3}$, to generate shock-mulitiplexing for increasing $\rho_{f,3}$, and to accelerate the shell up to 10^8 cm/sec for heating the fuel up to $5 \sim 10$ keV by the reflected shock wave. Such implosion mode is named "Stagnation-free implosion". The world record of neutron yield is attained by this mode using "LHART" target (Yamanaka *et al.*, 1986b, Takabe *et al.*, 1987).

5.4 Ignition and burn

A high pellet gain of about $Q = 10^2$ is required for realizing a reactor of ICF. This is because the driver efficiency η_d and coupling efficiency η_c

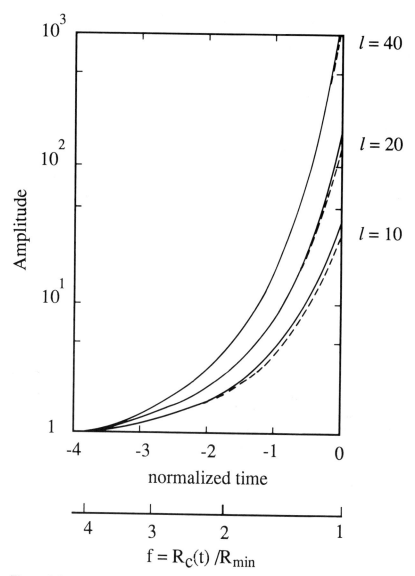

Figure 5.5 Temporal growth of perturbation amplitudes for $\ell = 10, 20$ and 40 transverse mode in stagnation phase. Solid and dotted lines show the WKB and numerical solutions. Time is normalized. $f = R_c(t)/R_{min}$ is convergence rate where R_{min} is the radius at maximum compression.

are limited to be $\eta_d \eta_c \ll 1$ and a core gain G = (thermonuclear energy)/(imploded DT fuel energy) in the order of 10^3 is necessary for a net gain. Assume that DT plasmas heated uniformly up to 10 keV, the core gain G of the thermonuclear reaction is calculated to be

$$G \approx 300f_B,$$

where f_B is the burn fraction. The core gain G cannot get the value 10^3 in a uniform heating. The high gain concept requires an ignitor scheme by which the compressed DT plasmas have an inner spark ignition with the temperature 5 ~ 10 keV and an outer main fuel region with the internal energy as low as possible. Then, the alpha particles produced in the spark region heat the main fuel to trigger the nuclear reaction. The ICF research is requested to demonstrate at first the ignition of the fuel and next to attain the high pellet gain.

5.4.1   Ignition condition

The ignition condition is given from the energy balance of the DT plasmas which are assumed to be uniformly heated and compressed. The energy equation to the plasmas with the assumption of $T_e = T_i = T$ is

$$3\frac{\rho}{m_i}\frac{dT}{dt} = -P\nabla u + \nabla K_e \nabla T - L_R + H_\alpha, \qquad (5.2.1)$$

where the first term of the right part gives the expansion loss L_E, the second one is thermal conduction loss L_C, the third one represents radiation cooling L_R, and the fourth one is the alpha-particle heating. If the DT plasmas have the radius R, density ρ, and temperature T, each cooling term of Eq. (5.2.1) is roughly evaluated to be

$$L_E = P\nabla u \approx 2\frac{\rho T}{m_i}\frac{3}{R}2C_s = 1.27 \times 10^{23}\frac{\rho}{R}T_k^{3/2}, \qquad (5.2.2)$$

$$L_C = -\nabla K_e \nabla T \approx K_0 T^{7/2}\frac{1}{R^2} = 8.7 \times 10^{19}\frac{1}{\ln\Lambda}\frac{T_k^{7/2}}{R^2}, \qquad (5.2.3)$$

$$L_R \approx A_0 \rho^2 T^{1/2} = 2.77 \times 10^{23}\rho^2 T_k^{1/2}, \qquad (5.2.4)$$

where each last formula is in cgs units and T_k is the temperature in keV units. In Eq. (5.2.2), we evaluated $\nabla u = -1/\rho\, d\rho/dt$ as $\rho \propto R^{-3}$ and $dR/dt \approx 2C_s$, where C_s is the sound velocity.

The alpha-particle heating term is given in the form;

$$H_\alpha = \left(\frac{\rho}{2m_i}\right)^2 <\sigma v>_{DT} f_\alpha \xi_\alpha = 8.07 \times 10^{40} <\sigma v>_{DT} f_\alpha \rho^2, \qquad (5.2.5)$$

where $\xi_\alpha = 3.52\,\mathrm{MeV}$, $f_\alpha < 1$ is the energy deposition fraction of the alpha-particle, and $<\sigma v>_{DT}$ is a function of only the temperature. The fraction f_α is roughly given as $f_\alpha \approx 1 - (1 - \rho R/\rho \lambda_\alpha)^2$ for $R < \lambda_\alpha$ and $f_\alpha \approx 1$ for $R \geq \lambda_\alpha$, where λ_α is the stopping range of the alpha-particle. Note that $\rho \lambda_\alpha$ slightly depends upon the density (Fraley et al., 1974) and approximately equal to $\rho \lambda_\alpha \approx 0.04\,T_k\,(g/cm^2)$ in the region of interest.

Then, we define the ignition condition by the fact that the right part of Eq. (5.2.1) balances each other; namely, $L_E + L_C + L_R = H_\alpha$. Multiplying R^2 to Eq. (5.2.1), we obtain the relation between ρR and T_k.

$$8.7 \times 10^{18} T_k^{7/2} + 1.27 \times 10^{23} T_k^{3/2} \rho R + 2.77 \times 10^{23} T_k^{1/2} (\rho R)^2$$
$$= 8.07 \times 10^{40} <\sigma v>_{DT} f_\alpha (\rho R)^2. \qquad (5.2.6)$$

This relation is shown in Figure 5.6 with thick solid line. For reference, other conditions obtained for $L_C + L_R = H_\alpha(CR)$, $L_E = H_\alpha(E)$, $L_C = H_\alpha(C)$ and $L_R = H_\alpha(R)$ are plotted. As seen in this figure, when the expansion loss is dominant, $\rho R \simeq 0.4\,g/cm^2$ at 10 keV is the ignition condition.

A number of simulations of the fusion burn have been performed by Fraley et al. (1974) with uniformly compressed DT fuel. As well-known, a simple estimate of the fusion burn fraction without taking into account the self-heating is (Bodner, 1981);

$$f_B = \frac{\rho R}{\rho R + \beta(T_i)}, \qquad (5.2.7)$$

where $\beta(T_i) = 8m_i C_s / <\sigma v>_{DT}$ is a function of only T_i. It is noted that for $\rho R \ll \beta(T_i)$, $f_B \approx \rho R/\beta(T_c)$ and the burn fraction increase proportionally to ρR. The ignition point is given by the ρR value at which the f_B departs to increase from the linear line in Figure 5.7.

5.4.2. Necessary energy for ignition

The DT plasmas with areal density ρR and temperature $T_i(= T_e)$ have internal energy U_{DT};

$$U_{DT} = 10.6 \times 10^9 \frac{1}{(\rho/\rho_s)^2} (\rho R)^3 T_k \,(\mathrm{Joule}) \qquad (5.2.8)$$

where ρ_s is the solid density and ρR is in g/cm^2. In Figure 5.8, three lines indicate $(\rho R)^3 T_k = 0.1$, 0.5, and 1.0, respectively. For a coupling efficiency of 10%, the driver energy of 10% kJ is enough for ignition defined by curve A. Therefore, if a spherically symmetric implosion was done, $U_{DT} = 1$ kJ is an ignition condition at $\rho/\rho_s = 10^3$ for $(\rho R)^3 T_k = 0.1$. However, the implosion symmetry is a key issue and the

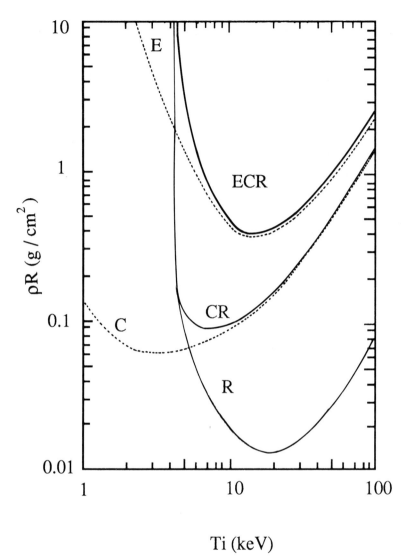

Figure 5.6 Ignition conditions evaluated from simple energy balances.

estimation of U_{DT} including the asymmetry effect is essential. The curve B in Figure 5.8 shows the 10 times severe ignition condition estimated from Figure 5.7. The dotted curve is estimated by a simple energy balance. If the core radius R is deformed with a perturbation with amplitude ξ due to the nonuniformity of laser irradiation or other effects; then, the effective ρR contributing to the fuel burn will be given by,

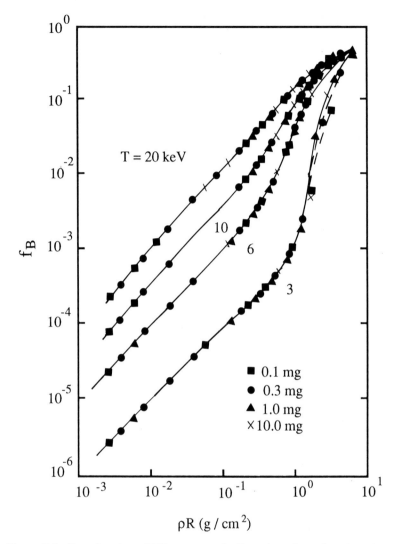

Figure 5.7   Burn fraction of DT core numerically evaluated as a function of core radius ρR with initial temperature 3, 6, 10 and 20 keV at various fuel mass.

$$(\rho R)_{eff} = \rho(R - \xi)$$

$$\approx \rho R(1 - \alpha\xi/R_0) \qquad (5.2.9)$$

where $\alpha = R_0/R \approx (A_0/3\rho/\rho_s)^{1/3}$ and R_0 and A_0 are the initial radius and the aspect ratio. Inserting ρR of Eq. (5.2.9) into Eq. (5.2.8), for the cases

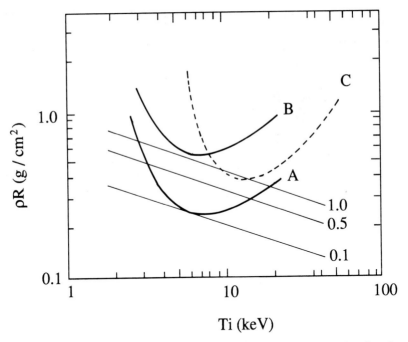

Figure 5.8 Ignition characteristics of DT fuel using the ignition point given by Figure 5.7. Curve A corresponds to 10% deviation from Eq. (5.2.7) and curve B for 100% deviation. Curve C is due to the simple energy balance. The relations of $(\rho R)^3 T = 0.1$, 0.5 and 1.0 are plotted by thin solid lines.

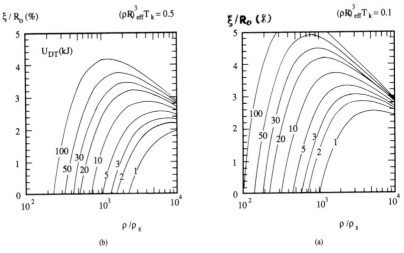

Figure 5.9 Core uniformity ξ/R_0 and energy requirements U_{DT} for ignition are given with various compression rates ξ/ξ_s.

of $(\rho R)^3_{eff} T_k = 0.1$ and 0.5 at $A_o = 10$, we get Figure 5.9 (a) and (b), respectively. In Figure 5.9, if we set $U_{DT} = 10$ kJ and $\rho/\rho_s \simeq 10^3$, the required uniformities are 3.5% for $(\rho R)^3_{eff} T_k = 0.1$ and 1.5% for $(\rho R)^3_{eff} T_k = 0.5$ respectively. Therefore, for a substantial ignition a uniformity of ξ/R_0 equal to $2 \approx 3\%$ is requested for a 100 kJ driver with the coupling efficiency 10%.

In the case of the indirect drive, the coupling efficiency is less than 1%, however the uniformity of the core is much better than that of the direct drive. The required driver energy is expected to be a few MJ. The choice of the direct drive and the indirect drive is a controversial issue for ICF research.

6 EXPERIMENTAL STATUS

In the 1970s the laser plasma interaction was one of the most important research items using available lasers of a modest size. These experiments on laser plasmas gave us the basic understanding of laser coupling to plasmas. In the 1980s several huge lasers enabled us to perform the implosion experiments. For the ICF, important issues to investigate accompanied by several items are shown as follows:

(1) Absorption mechanism: plasma waves, saturation of instability, electron temperature, ion temperature.
(2) Energy transportation: density gradient, electron temperature, X-ray radiation temperature, flux limiting factor.
(3) Implosion process: shock velocity, uniformity, mass ablation rate, density, areal density, core size, temperature.
(4) Ignition and burn: neutron, α particle, reaction rate, energy yield.
(5) Energy extraction: energy balance, efficiency.

To get the information on these items, we have performed several experiments on the fundamental properties of the laser plasma.

6.1 Interaction experiments

In Figure 4.4, the absorption properties of a plasma due to the different wavelength of lasers are shown. The anomalous absorption appears beyond the range of $I\lambda^2 \approx 10^{14}\,\mathrm{W/cm^2 \cdot \mu m^2}$ where I ($\mathrm{W/cm^2}$) is the laser intensity and λ is the laser wavelength. Anomalous absorption has a tendency to produce hot electrons due to the nonlinear process of coupling which is easily introduced by the longer wavelength laser. As well known, the hot electrons induce the preheat of the fuel which prevents the efficient compression of the fuel pellet. This is a reason why the short wavelength lasers become important instead of the highly efficient CO_2 laser. The classical absorption is dominated by the shorter wavelength laser light.

In Figures 5.2 and 6.1, the ablation pressure due to the laser absorption and the compressed fuel density versus the ion temperature are given comparing the effect of the wavelength of lasers respectively. These results also indicate the importance of the shorter wavelength laser.

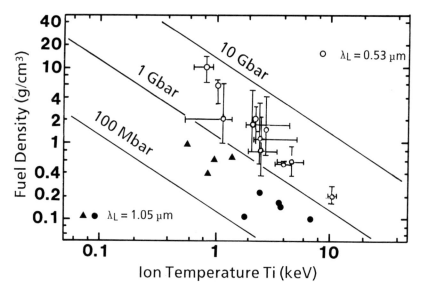

Figure 6.1   Compressed fuel density versus ion temperature by laser irradiation.

6.2   Direct driven experiments

The direct implosion scheme is so efficient at ablating the target that the modest size laser is available for the experiment. The key issue for this scheme is to attain uniform compression which needs uniform irradiation of laser as well as high uniformity of the fuel pellet. To eliminate the nonuniform irradiation due to the diffraction of the coherent laser light, several new methods to suppress the coherency of laser light are introduced such as the random phase plate, the induced spatial incoherence grating with a broadband oscillator and the multi-lens irradiation concept.

As for the pellet uniformity, super high uniform pellets are prepared by glass or plastic microballoons. The cryogenic DT target is also under investigation for this purpose. Several types of target are shown in Figure 6.2.

6.2.1   High neutron yield target

Recently we have succeeded in performing a high neutron yield experiment using a novel target LHART: Large High Aspect Ratio Target. This target is a glass microballoon with an extremely thin shell which contains a fuel gas. When it is irradiated by a long laser pulse, almost all of the pusher layer is exploded until it reaches the maximum compressed condition stage. Then, the stagnation free compression is introduced

Target	η_{abs}	η_c
Ablative target 	50 % Higher efficiency by blue High uniformity requirement	1 %
LHART 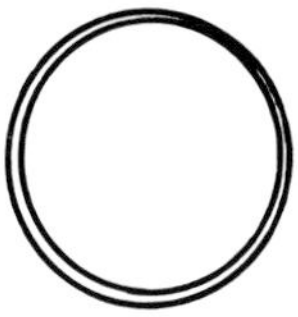	70 % Higher efficiency by blue Stagnation free	6 %
Double shell 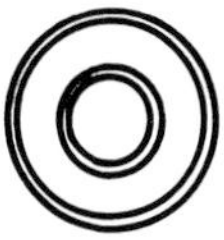	50 % Uniformity improved Velocity multiplexing	2 %
Cannon ball 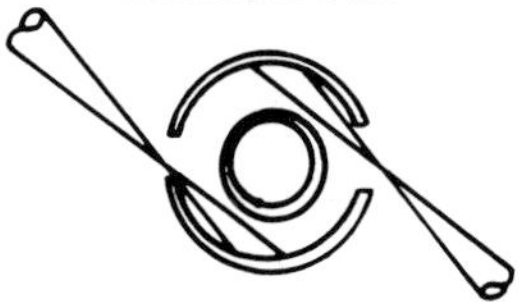	70 % High efficiency Uniformity improved	6 %
Double shell in cannon 	70 % High efficiency Uniformity greatly improved	1 %

Figure 6.2 Several types of laser fusion target. Laser absorption η_{abs} and core coupling efficiency η_c are given.

without the pusher-fuel mixing which is introduced by the Rayleigh-Taylor instability.

In Figure 6.3 a schematic diagram of the ablative mode compression for a thick shell (low aspect ratio) target is shown. By the ablative implosion, the shock wave is firstly driven into the fuel gas and secondly

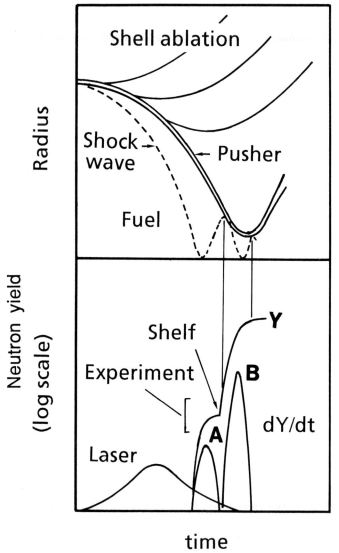

Figure 6.3 Schematic diagram of ablative mode compression.

the pusher follows toward the center of the pellet. According to our simulation the neutron yield is observed twice, the first one is due to the shock arrival to the center indicated by A and the second one is due to the adiabatic compression indicated by B in Figure 6.3. However the actual experimental fact shows the single peak of the neutron yield. This result is explained by the disappearance of the second neutron pulse due to the cooling of the fuel which is caused by mixing of the pusher and the fuel during the stagnation.

From this understanding, to get a high neutron yield we can conclude that the stagnation free compression is preferable to coincide with the two peaks of neutron yield. This scheme was realized by using 'LHART'. In 1986, the D-T neutron yield of 10^{13} was attained by the green beam 13 kJ, 1 ns, irradiation of GEKKO XII glass laser to the glass microballoon, diameter 1.235 μm, shell thickness 1.31 μm, aspect ratio 471 and enclosed DT fuel 6.2 atm. The pellet gain reached 1/500. Figure 6.4 shows the neutron yield versus the pellet aspect ratio. From the observation of X-ray pinhole pictures of pellet implosion, a ring type image of the compressed core produces larger neutron yield than the case of a solid core image. The largest neutron yield is attained at the aspect ratio 400 ~ 500. (Yamanaka, 1986b)

For the case of the aspect ratio smaller than 400, well known ablative compression is introduced where Rayleigh-Taylor instability becomes dominant to induce the fuel-pusher mixing and to suppress the neutron yield. When the aspect ratio is larger than 500, the shell pellet is thin enough to introduce the burn through of the pusher which prevents the uniform compression. In the case of an aspect ratio of about 450, the stagnation free compression is introduced to get the maximum neutron yield which accords very well with the 1D simulation data given by the ILESTA code.

6.2.2 High compressed density target

A gas-filled target has some limitation to attain the high compressed density. Shell fuel targets such as cryogenic DT shells and plastic CDT shells are very interesting. We have imploded super uniform plastic fuel CDT shells to get the high compressed fuel density. The target has a diameter of 500 μm, shell thickness 6 ~ 12 μm, tritium content > 500 Ci/cm^3 and 5 wt % Si added which are assumed constant during the implosion. The green beam 8.5 kJ, pulse duration 2 ns from GEKKO XII laser with random phase plates is irradiated uniformly. After irradiation we can estimate the plasma ρR value by the silicon activation method which was reported recently. In this case the activation reaction ^{28}Si(n, p)^{28}Al is used.

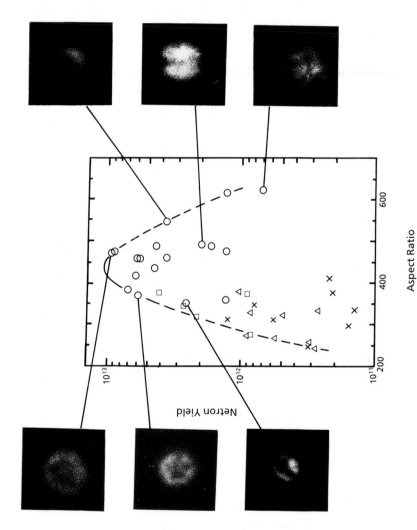

Figure 6.4 Neutron yield versus pellet aspect ratio. X-ray pinhole pictures are also shown.

As well known, the activation is proportional to the neutron yield and plasma ρR. At the same time, multilayer labelled shells with a coating of Mg, Si, Al can show the ablated mass quantity ΔM by the X-ray streak photography. From the experiment we have data on the neutron yield, the Si activation, ρR and mass ablation ΔM for imploding polymer shells which allow us the final fuel density ρ (Mima, 1988).

The measured minimum density ρ_{fuel} is $89\,\mathrm{gcm^{-3}}$ that is about 600 times the liquid fuel density. These data are given in Figure 6.5. The final goal of density is believed to be 1000 times the fuel liquid density. The

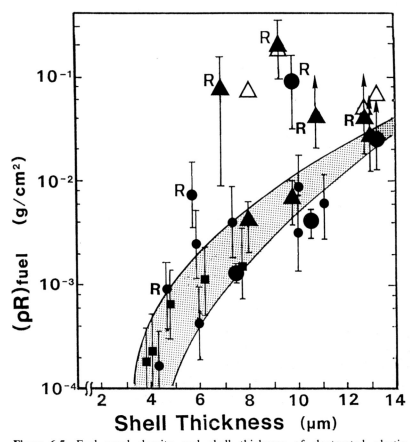

Figure 6.5 Fuel areal density and shell thickness of deuterated plastic shell targets. The shaded area corresponds to the normal laser beam and R indicates the data for the random phasing beam. $\triangle$ symbols show the data of 1D simulation.

situation is now very encouraging. Experiments on the cryogenic fuel target are highly expected.

6.3   Indirect driven experiments

Indirect drive is an approach to ICF that relaxes the requirements of implosion uniformity for the direct drive and simplifies the laser configuration of irradiation. There are two categories of indirect driven target: the plasma driven cannonball target and the radiation driven target. The cannonball target is fundamentally a double shell structure, the cavity of which is filled by the plasma or the X-ray which is produced by the impinging laser. These targets are shown in Figure 4.4.

The hydrodynamic efficiency of the plasma driven cannonball is proportional to cube of the aspect ratio ρ (ratio of inner and outer radius), so a low aspect target has a poor performance. The inlet hole of laser beams into the cavity is gradually closed by plasma. In contrast the radiation driven cannonball target has better performance owing to the fact that the absorption of X-rays on the inner fuel pellet is proportional to the square of the aspect ratio ρ. And also the laser light irradiates in two ways: one is through the inlet hole into the cavity and the other is to irradiate directly the outer shell where the produced X-ray penetrates into the cavity. The important issues for cannonball targets are the X-ray conversion efficiency from the laser light and the preheat problem of the inner pusher by X-rays.

The cannonball target has several advantages such as good uniformity, less sensitivity to the laser wavelength and also to the laser irradiation configuration. However the driver power for the cannonball target should be larger than that for the direct drive due to the smaller coupling efficiency. The LLNL is also using a similar concept of the indirect drive named by an old German term "Hohlraum".

6.3.1   X-ray conversion

The converted X-ray from the high density and high temperature plasma produced by lasers is very bright in the soft X-ray range of $100 \sim 300$ eV. The experimental and theoretical investigations of the soft X-ray by lasers have been developed.

The ionization equilibrium time τ in the high density plasma is given by

$$\tau \sim \frac{4.5 \times 10^7 Z^3}{n^4 N_e} \left(\frac{kT}{Z^2 E_H} \right)^{1/2} \exp\left(\frac{2Z^2 E_H}{n^3 k T_e} \right)$$

where T_e and N_e are the temperature and the density of electrons, n is the principal quantum number of the orbital electron of ions and E_H is 13.6 eV. If we put $kT_e \sim 1$ keV, $n = 1$, $Z \sim 10$, τ becomes $10^{12}/N_e$. The ionization state of plasma of $N_e \sim 10^{23}$ cm^{-3} will reach the equlibrium in 10 ps. The local thermoequilibrium is likely to be established.

However in a laser produced plasma, the transient properties due to the larger density gradient and also the plasma fluid motion are observed. During the laser heating, the high energy radiation due to the bound to bound level transition is emitted from the low density, high temperature corona. And the low energy radiation is dominant from the high density, low temperature region of ablation by the free to bound transition. In a high Z matter, such as gold, multi-ionized states are introduced where the number of states of orbital electrons is so large that the many radiation lines seem to form a quasi-continuum. However they are far from the black-body radiation in the case of a plane target plasma.

In Figure 6.6, the converted X-ray radiation spectrum due to the laser irradiation versus atomic number Z is given. The conversion efficiency

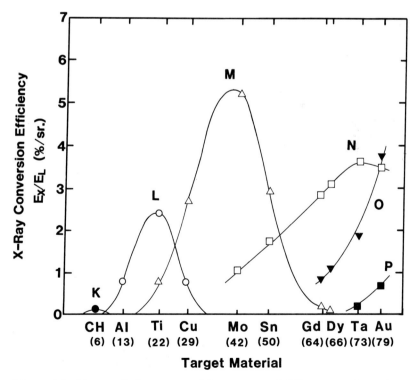

Figure 6.6   X-ray radiation converted from targets of different atomic number.

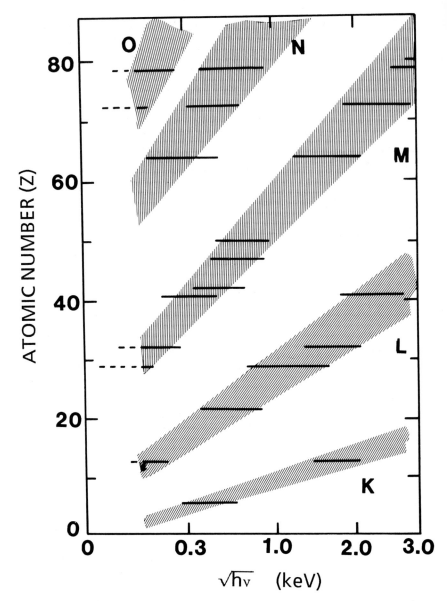

Figure 6.7 X-ray radiation spectrum corresponding to electron orbital transitions in K, L, M, N and O shells of ions.

increases with the increase of atomic number Z and has some undulation. The X-ray radiation spectrum has several humps which corresponds to the electron orbital transitions of the ion such as K, L, M, N and O shells. They show a similar relation of the classical Mosley law which indicates a strong emission at $Z \propto \sqrt{h\nu}$. These relation is indicated in Figure 6.7. (Mochizuki, *et al.*, 1986)

In a cavity target plasma such as the cannonball target, the laser produced X-ray is absorbed in a cavity and reemitted. The radiation confinement in a cavity can keep a longer time of X-ray emission. Figure 6.8 shows the experimental results done by the joint work of ILE Osaka and MPQ Garching which indicates the X-ray intensification in the cavity. (Mochizuki, *et al.*, 1987)

6.3.2 Cannonball target experiments

A cannonball target is a double shell target in which laser energy is injected into a cavity formed between the two shells. The outer shell acts as an energy container, whereas the inner shell acts as a pusher for compression of the fuel. The laser energy is introduced into the cavity through the holes in the outer shell or through the outer shell itself. The energy of the laser beams is converted to the plasma kinetic energy and the radiation energy inside the cavity resulting in uniform compression of the fuel. It is the reason why this target is named as a cannonball target.

As mentioned before, there are two modes of compression for the cannonball target:plasma driven cannonball and radiation driven cannonball. When the cavity size is small, it is filled with high temperature plasmas which compress the inner shell by the high plasma pressure. In this mode, the laser pulse has to be short enough to inject the laser energy effectively into the small cavity against the inlet hole closer due to the plasma formation. This scehme needs the high intensity irradiation which introduces the preheating of the fuel by high temperture electrons produced in the cavity.

Nowadays, the radiation cannonball scheme is dominated those cavity size is rather large irradiated by the longer laser pulse. The laser energy can be efficiently converted to the soft x-ray which ablates the inner shell to implode the fuel. The very uniform compression is expected in this scheme by using the shorter wavelength laser light.

As an example, the behaviors of the 2 holes cannonball targes irradiated by the $0.52 \, \mu m$ laser light are presented.

(1) Mass ablation of the inner shell
In Figure 6.9, the x-ray streak image of a cannonball targert experiment

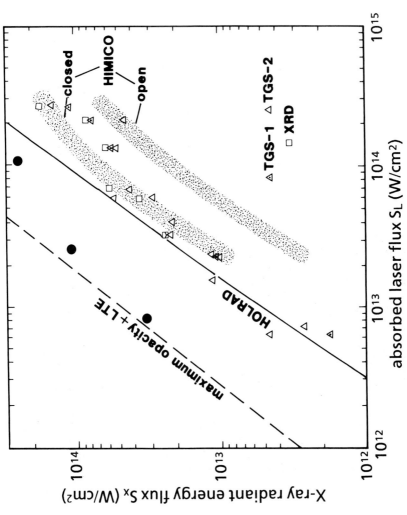

Figure 6.8 Radiant energy flux S_x through a diagnostic hole of a laser heated cavity versus averaged absorbed laser flux S_L. TGS: transmission grating spectrum, XRD: X-ray diode signal. The open symbols are the data with green laser pulse 0.3 ns. The black circles are the recent results by the blue laser of 0.8 ns pulse. Several simulation results are also indicated.

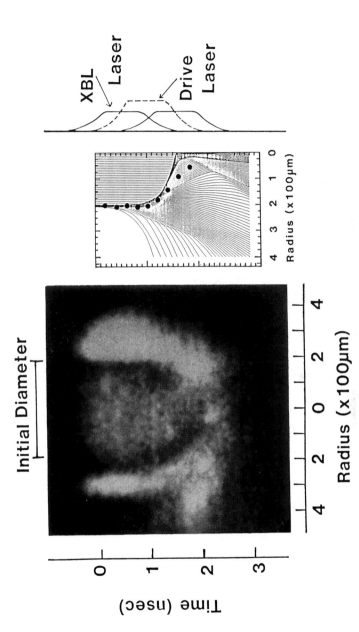

Figure 6.9 Typical X-ray streaked shadow image of the inner shell of a cannonball target and the relevant simulated flow diagram. The closed circles represent the density peaks derived from the shadow data using the Abel-inversion. The outer shell of the cannonball was a 2000 μm diam. gold shell, and the inner shell was a multilayer 2.7 μm coated glass microballoon of 401 μm diameter and 0.98 μm thickness.

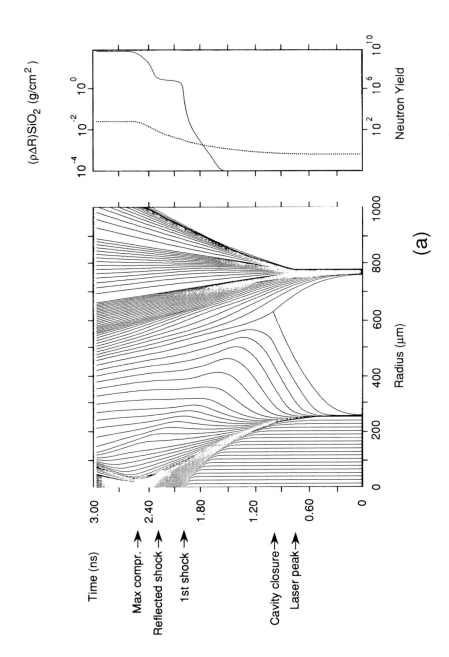

(a)

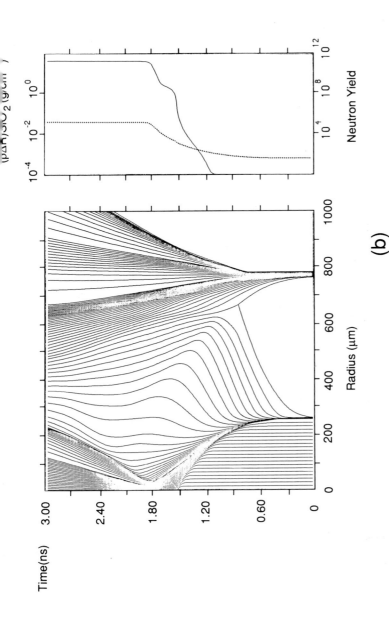

Figure 6.10 1-D hydrodynamic simulation of an outer diameter 1560 μm/inner diameter 500 μm two hole cannonball target irradiated by 2ω/500ps/4800J laser. Hydrodynamic motion and the time-integrated neutron yield are shown. In the upper figure the absorbed laser intensity on the inner shell was the same as that on the outer shell, whereas in the lower figure the former was three times higher than the latter.

whose outer shell diameter $2000\,\mu$m, inner shell diameter $500\,\mu$m irradiated through two holes by the $0.52\,\mu$m wavelength, 500 ps pulse duration, 5 kJ laser. The SiO_2 inner shell $0.98\,\mu$m thick is coated with $0.4\,\mu$m thick C_8H_8, $0.8\,\mu$m Al and $1.5\,\mu$m C_8H_8 from the inside to outside respectively. From the characteristic X-ray emission, the mass ablation rate $\dot{m}$ is determined as 8.6×10^5 g/cm^{2}s. The mass ablation rate by X-ray irradiation is given by

$$\dot{m} = 1.5 \times 10^5 I_R^{1/3} \lambda^{-4/3} (A/2Z)^{2/3}, (g/cm^2 \cdot s)$$

For $T_R = 130$ eV, $I_R = 2.7 \times 10^{13}$ W/cm^2, $A = 12$ and $Z = 6$ for CH layer we get m $\sim 2.3 \times 10^6$ g/cm^2 s which is near the observed value. (Fabbro $et\,al.$, 1985)

(2) Implosion dynamics

The X-ray streak image of Figure 6.9 is compared with the simulation results done by one-dimensional hydrodynamic code coupled with a non-LTE radiation multi-group transport package. In Figure 6.10 the hydrodynamic motion and the time-integrated neutron yield by 1D hydrodynamic simulation are given. The implosion speed is very close to the experiment. The compressed areal density of the pusher measured by the neutron activation of Si is 5.5 mg/cm^2 and the neutron is 1.6×10^9. The observed neutron yield is consistent with the yield at the first shock convergence in the simulation but is inconsistent with that at the maximum compression stage (Kato $et\,al.$, 1985).

The indirect driven target experiments have also been performed at Livermore where quite a lot of information is available. In this scheme, a more powerful laser or driver is essentially important. The heavy ion beam fusion is an interesting driver in this field.

7 DIAGNOSTICS OF IMPLOSION

The most important issues of plasma diagnostics are measurements of density and temperature of the imploded fusion fuel. These measurements have been remarkably developed in the last few years.

7.1 Density measurements

The measurements using nuclear fusion reaction products are very effective. We can use three methods to measure the fuel areal density $\langle \rho R \rangle$.

(i)   Neutron activation method
(ii) Knock on method
(iii) Secondary nuclear fusion reaction method

Since the thermonuclear burn time is short compared with the implosion time, measurements of the fuel density by reaction products are very reliable methods.

7.1.1 Neutron activation method

Neutron induced activation of Si which is seeded in the pusher of a pellet has been developed (Lane *et al.*, 1980). ILE has performed an experiment to give the compressed core density using the 14 MeV neutron from D-T reaction for the activation method to detect the areal density $\langle \rho R \rangle$ which is combined with the mass ablation measurement by using the streak photography of the implosion process.

The activation reaction is ^{28}Si(n, p)^{28}Al. The activation N is proportional to the neutron yield Yn and pusher ρR. The neutron yield is expected to be more than 10^{10} for a reliable measurement of ρR. The shell targets of CDT plastics, diameter $500 \sim 700\,\mu m$, shell thickness $5 \sim 15\,\mu m$ doped with Si in 5.2 weight percentage are irradiated by the second harmonic green light of the GEKKO XII glass laser, pulse duration $1.5 \sim 2\,ns$, energy $10\,kJ$ using the random phase plate to keep the uniform irradiation.

One must collect the activated elements which are ablated during the laser irradiation. Using a collective umbrella surrounding the fuel target, the collection factor of the activated ^{28}Al is precisely measured by a different experiment for the calibration. In this measurement the constant concentration of Si, C, D and T is assumed in the process of implosion.

Figure 6.5 is the experimental data for the areal density $\langle \rho R \rangle$ and the shell thickness of plastics CDT targets.

As for the measurement of ablated mass ΔM, one can use the characteristic X-ray emissions from multilayer labelled shell targets as shown in Figure 7.1. (Yamanaka *et al.*, 1987a) The data on $\langle \rho R \rangle$ and ΔM allow us to determine the final compressed core density ρ_{fuel} by the next relation.

$$\rho_{fuel} = \sqrt{\frac{4\pi/3 \langle \rho_{fuel} R \rangle^3}{(M - \Delta M)_{fuel}}}$$

For instance the experiment gives $(M - \Delta M)_{fuel} = 0.5\,\mu g$ and $\langle \rho_{fuel} R \rangle = 0.1\,g/cm^2$. Then the compressed core density is estimated as $\rho_{fuel} = 90\,g/cm^3$ that is 600 times the liquid fuel density. (Mima, 1988)

Another activation measurement is to seed a tracer gas such as ^{80}Kr into the DT fuel in a glass microballoon (Prussin *et al.*, 1986). The tracer is activated by $^{80}Kr(n, 2n)^{79m} Kr$. The ratio of the activation N^* to the neutron yield Y_n is given by

$$\frac{N^*}{Yn} = f \frac{\sigma \beta \langle \rho R \rangle_{tracer}}{m_{tracer}}$$

where σ is the cross-section of the activation, β is the isotope abundance, $\langle \rho R \rangle_{tracer}$ and m_{tracer} are the areal density and the atomic mass of the tracer respectively. The factor f is a parameter dependent on the profile of the burn region which is one when the profile is symmetrical.

The neutron activation method can be applicable for the $\langle \rho R \rangle$ value greater than the ignition condition, $0.2 \sim 0.3\,g/cm^2$.

7.1.2 Knock on method

For the knock on method it is not necessary to use a tracer like the neutron activation method. The number of elastically scattered deuterons and tritons, Y_{ko} is given by

$$Y_{ko} = f \frac{\sigma_D + \sigma_T}{m_D + m_T} \langle \rho R \rangle Yn$$

where $\sigma_D(\sigma_T)$ and $m_D(m_T)$ are the elastically scattering cross section and the mass of D(T) respectively. The Y_{ko} is detected by the tracks in CR 39 films. The track detector has been developed to count automatically the track numbers in a film. However, this method is limited to a target areal density below $0.07\,g/cm^2$ due to the thermalization of the scattered particles through the higher density fuel (Kacenjar *et al.*, 1984).

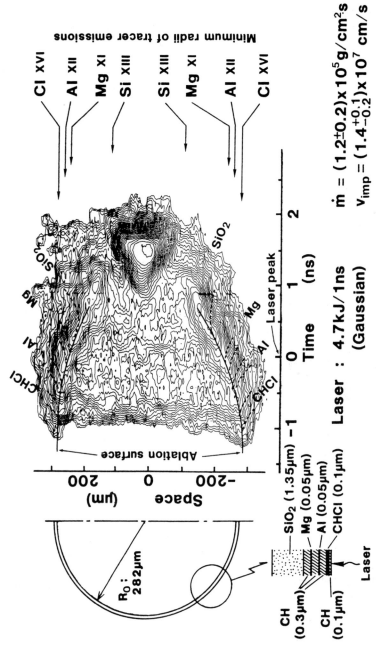

Figure 7.1 X-ray emissions from multilayer labelled shell targets.

7.1.3 Secondary fusion reaction method

This method does not depend upon the tracers but needs a high neutron yield. Secondary nuclear fusion reactions in deuterium fuel are given in Figure 7.2. The numbers of DT and D^3He reactions are measured as the yields of secondary DT reaction neutrons, Y_{2n}, and secondary D^3He reaction protons, Y_{2p}, respectively. The yield ratios of Y_{2n} and Y_{2p} to the primary neutron yield Y_{1n} are given by

$$\frac{Y_{2n}}{Y_{1n}} = f\frac{\sigma_{DT}<\rho R>}{m_D}$$

$$\frac{Y_{2p}}{Y_{1n}} = f\frac{\sigma_{DHe}<\rho R>}{m_D}$$

where $\sigma_{DT}(\sigma_{D^3He})$ is the cross-section of the DT(D^3He) reaction averaged over the path of the tritons (^{3}He nuclei). Results of secondary DT neutron time-of-flight measurements are shown in Figure 7.3. DT neutron signals (peak b) are clearly seen. In a fuel having a high $\langle\rho R\rangle$ value and a low electron temperature, the relations between $\langle\rho R\rangle$ and these yield ratios (Y_{2n}/Y_{1n} and Y_{2p}/Y_{1n}), become complex. In this case, simultaneous measurements of the secondary neutrons and protons or neutron spectra measurements are required (Azechi *et al.*, 1986; Bazov *et al.*,

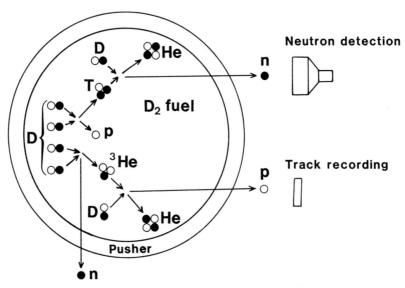

Figure 7.2 Secondary nuclear fusion reaction of DT in DD plasma.

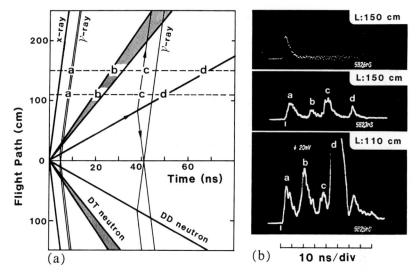

Figure 7.3 The time of flight measurement of secondary DT neutrons.

1986; Cable *et al.*, 1986; Gamalii *et al.*, 1975). The method using secondary fusion reactions is limited to a fuel areal density below 0.1 g/cm² at an electron temperature of 3 keV due to the thermalization of the 1.01 MeV tritons in the fuel.

In Figure 7.4 the regions in which each of the three methods to determine $\langle \rho R \rangle$ can be applied are schematically illustrated. The lower boundaries are set by the sensitivity of measurements. Upper boundaries of the knock-on and the secondary reaction methods are imposed by the stopping of charged particles in the fuel.

7.2 Temperature measurements

The ion temperature is measured using Doppler broadening of fusion products. Neutrons are suitable for this method since the velocity change in the target is negligible. A time-of-flight (TOF) spectrometer is commonly used because of the short burst and small size of the thermo-nuclear burn in targets (Lerche *et al.*, 1977). For the flight distance L and the time width Δt (full width at half maximum), the fuel ion temperature T_i is given by

$$T_i(\text{keV}) = 68.5[\nabla t(\text{ns})/L(\text{m})]^2$$

At low ion temperature (< 1 keV), difficulty arises in this method due

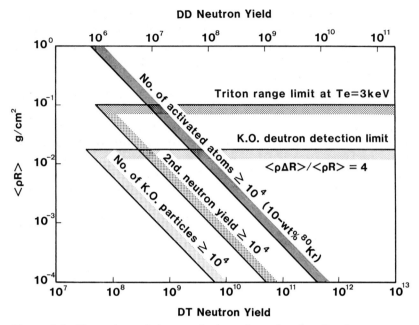

Figure 7.4 The regions of three methods to determine the pR value.

to the low neutron yield and the short time width. Reaction ratio measurements, such as the ratio of DD to DT reactions, is one of the solutions in the low-temperature region.

7.3 Image processing

The compression uniformity is measured by two- or three-dimensional imaging of thermonuclear reaction products. Since pinhole imaging requires a high yield of the reaction products, various coded aperture imaging (CAI) processes have been proposed to increase the solid angle of observation. Three CAIs have been demonstrated: Fresnel zone plate (FZP); uniformly redundant array (URA); and penumbral imaging.

The experimental demonstration of α-particle imaging with a FZP camera having a resolution of 5–10 μm has been attained (Ceglio *et al.*, 1977). An image through the FZP illuminated by α-particles is constructed on a track detector as an encoded image. The particle source image is optically reconstructed by irradiating the encoded image by a laser light.

A X-ray imaging using a URA camera has been demonstrated (Yamada *et al.*, 1983). In contrast to FZP imaging, numerical reconstruction of the URA coded image is rather straightforward. The nonlinearity of the track detector is easily taken into account. A key problem in numerical reconstruction in CAI is the broadening of a system point spread function (SPSF). Since the SPSF of URA images is a δ function, artifact-free reconstruction is possible.

To resolve the future high density compressions, it is necessary to use particles of a long range, such as neutrons or 14.7 MeV protons by D^3He reactions. There are technical difficulties in making fine apertures of 10 μm in size on a plate with thickness of the order of the ranges of these particles. Penumbral imaging has been proposed to solve this problem (Nugnet *et al.*, 1984). The necessary aperture is just a circle that is larger in size than the compressed core. A penumbra is formed at the edge of the shadow of the aperture irradiated by particles from the compressed core as shown in Figure 7.5. The reconstruction is made with the spatial derivative of the penumbra.

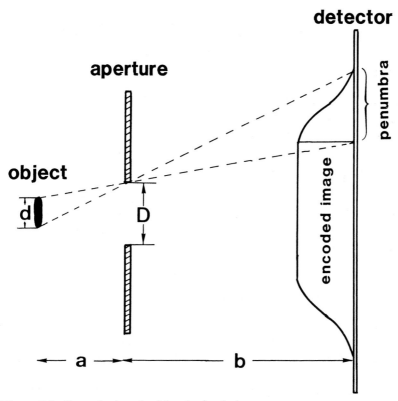

Figure 7.5 Penumbral method for the implosion core.

Obviously three-dimensional imaging is a more useful technique for the study of compression uniformity. Recently, a computer tomography of an X-ray image using URA cameras has been demonstrated (Chen et al., 1986).

7.4   X-ray measurements

According to the emission spectrum of X-ray, the spectroscopic measurements are divided into three regions. The first part is the range of sub keV X-ray measurement. The laser irradiated hi Z matter, such as Au, emits X-rays of this range with a high conversion efficiency of $20 \sim 80\%$. This emission is very effective for the indirect drive implosion using a radiation cannonball target. The absolute measurement of time resolved X-ray spectrum is fundamentally important for the implosion research.

The second part is the spectrum range of $1 \sim 10\,\mathrm{keV}$. They are closely related to the resonance lines and the recombination lines of the target materials. The information of the plasma parameter around the cut-off density region is available from their spectra.

The third part is the range of $10 \sim 100\,\mathrm{keV}$ light, which is emitted from the corona region to the laser absorbed region where the hot electrons are generated by the laser-plasma interaction. They are very important to understand the implosion properties.

7.4.1   X-ray spectrum of laser plasma

As for the radiation process of high temperature and high density plasmas, there are two categories, recombination radiation due to the electron capture by an ion in plasma (free-bound transition) and Bremsstrahlung owing to the collisions between electrons and ions (free-free transition).

(1) Recombination radiation
When a free electron is captured by an ion, the sum of the initial kinetic energy and the ionization energy related to the capture-orbit is emitted ($h\nu = 1/2mv^2 + X_n$). For an hydrogen-like ion of the i th charge state, the photon energy per a unit emitted energy in the recombination ($N_{i+1} + e \rightarrow N_i, n + h\nu$) is given by

$$\frac{dE_{bf}}{dh\nu} = 1.7 \times 10^{-14} N_e, N_i + 1\, T_e^{-3/2} X_{i,n}^2 \left(\frac{\zeta_n}{n}\right) g_{bf}$$

$$\times \exp\left[\frac{-(h\nu - X_{i,n})}{T_e}\right] \ (\mathrm{keV/keV/s/cm^3/str}) \qquad (7.4.1)$$

Where N_e is the electron density, $N_i + 1$ is the ion density of $i + 1$ charge state, $X_{i,n}$ is the ionization potential from the final state, ζ_n is the vacant number of the n shell for captured electrons and g_{bf} is gaunt factor of free-bound transition. The spectrum of recombination radiation is given by the total sum of all ion species and electronic shells (Johnson, 1972; Cuderman *et al.*, 1972).

(2) Bremsstrahlung of dipole model approximation
A free electron is decelerated by the ion to produce a photon. The Bremsstrahlung radiation energy produced by electrons (N_e cm^{-3}) of temperature T_e and ions (N_i cm^{-3}) of Z_i charge state are

$$\frac{dE_{ff}}{dh\nu} = 2.4 \times 10^{-16} N_e N_i Z_i^2 T^{-1/2} g_{ff} \exp\left(\frac{-h\nu}{kT}\right), \ (\mathrm{keV/keV/s/cm^3/str})$$
$$(7.4.2)$$

where g_{ff} is the gaunt factor of free-free transition. Actual emissions are given by the total sum of all ion species to be approximately

$$\sum N_i Z_i^2 \sim N < Z^2 >.$$

According to the increase of the electron temperature, the ion is fully ionized and the Bremsstrahlung radiation will exceed the recombination radiation. The threshold for this critical point is $T_e > 4.08 \times 10^{-2} Z^2$ (keV) for Z-plasma (Hadlestone *et al.*, 1965).

(3) Bremsstrahlung of multipole model
When the electron temperature becomes larger than 30 keV, the dipole model for the interaction between electrons and ions fails to approximate the situation. Taking into the relativistic effect, the quadru-pole approximation is effective in this temperature region (Maxon, 1972).

The radiation energy is given by the sum of spectra due to the electron-electron interaction and the electron-ion interaction. In the case of $N_e = ZN_i$ and $Z = 1$, one can have

$$\frac{dE_{total}}{dh\nu} = \frac{dE_{ei}}{dh\nu} + \frac{dE_{ee}}{dh\nu}$$

$$\frac{dE_{ei}}{dh\nu} = \frac{16}{3}\left(\frac{2}{\pi m T_e}\right)^{1/2} ar_o^2 n_e^2 mc^2 \exp\left(-\frac{\lambda}{2}\right) K_o\left(\frac{\lambda}{2}\right)\frac{1}{h\nu}$$

$$\frac{dE_{ee}}{dh\nu} = \frac{T_e}{mc^2} B(\lambda)\frac{dE_{ei}}{dh\nu}$$

where $\lambda = \dfrac{h\nu}{T_e}, a = \dfrac{e^2}{2hc\epsilon_o}, r_o = \dfrac{e^2}{4\pi c^2 m\epsilon_o},$

Ko(X) is the second kind of modified Bessel function,

$$Ko(X) \doteq e^{-x}\left(\frac{1.57}{x}\right)^{1/2}\left(1 - \frac{0.125}{x} + \frac{0.07}{x^2}\right)$$

and $B(\lambda)$ is the Born cross-section for the electron-electron Bremsstrahlung,

$$B(\lambda) \doteq 0.85 + 1.35\sqrt{\lambda} + 0.38\lambda$$

(4) Other X-ray radiations

Besides recombination radiation and Bremsstrahlung radiation there are two other forms of X-ray radiation. Line radiation occurs due to inner shell transitions in an exact analogy to the optical line emission. They are discussed in Section 6.

Another form of recombination can produce a significant flux of soft X-rays which is called dielectronic recombination. In this process an electron from the continuum recombines with an ion in a two-step process. The excess energy is transferred to another electron in the partially ionized ions. Then they decay to a ground state emitting photons. These photons tend to be very soft ($< 2\,keV$).

7.4.2 X-ray spectroscopy

In X-ray spectroscopy, self-absorption of the emissions is a serious problem. One approach for solving this problem is to use results of rate equation calculations in conjunction with hydrodynamic code calculations for the interpretation of experimental data (Duston *et al.*, 1983). Another way is to use line emission having low opacity or spectral shift.

Useful lines for measuring the electron temperature are satellite lines where upper levels are autoionization or dielectronic capture states and lower levels are excited states for He-like ions (Gabriel, 1972). Since the upper levels with ionic charge Z are populated from the ground state of ions with charge $Z + 1$, the intensity ratio of the satellite lines with different charges is a simple function of the electron temperature.

Useful lines for measuring electron density are the density-sensitive satellite lines (e.g. $2p^{2\ 3p} - 1s2p^{3p}$) (Presenyakav, 1976). As for the excitation to the upper levels of these transitions, autoionization levels are dominated by the electron impact transfer from adjacent autoionization levels (e.g. $2s2p^{3P}$), and hence the line ratio of these satellite lines to the other satellite lines ($2s2p^{3P} - 1s2s^{3S}$) is density sensitive. Since the rate of the electron impact transfer has a weak dependence on the electron temperature, some correction is necessary to calculate the exact density value.

Spectral shifts are caused by the screening of nuclear charges by free electrons and neighboring ions. The recombination edge shift is larger than the line shift. It can be used to measure the density for a medium compression case. Figure 7.6 shows edge shifts of recombination radiation ($Ne^{+10} + e \rightarrow Ne^{+9} + h\nu$) from imploded neon core. In Figure 7.6 (b) the density of the compressed neon was estimated to be $4.4\,g/cm^3$. With higher density and lower temperature compression, line shifts will be more useful as the line shifts become larger and the recombination radiation becomes weaker. Detailed theoretical treatments of level shifts have been discussed by several authors (Skupsky 1980; Yamamoto *et al.*, 1983; Kishimoto *et al.*, 1983).

X-ray shadowgraphy with a separate laser-produced plasma as an X-ray source is a useful tool to diagnose implosion dynamics (Key *et al.*, 1978). This technique can produce temporally frozen shadow images by means of a pulsed (100 ps) X-ray source.

A point X-ray source is useful to get a high photon energy with a laser facility of reasonable size. Point X-ray shadowgraphy has been

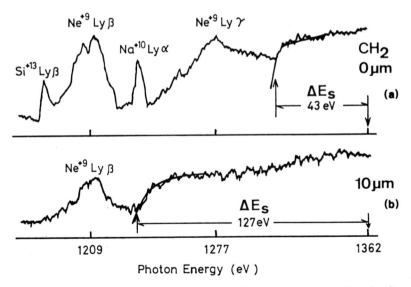

Figure 7.6 X-ray edge shift from imploded neon core gives the density $0.3\,g/cm^3$(a) and $4.4\,g/cm^3$(b). Ablater thickness CH_2 is $0\,\mu m$ (a) and $10\,\mu m$ (b).

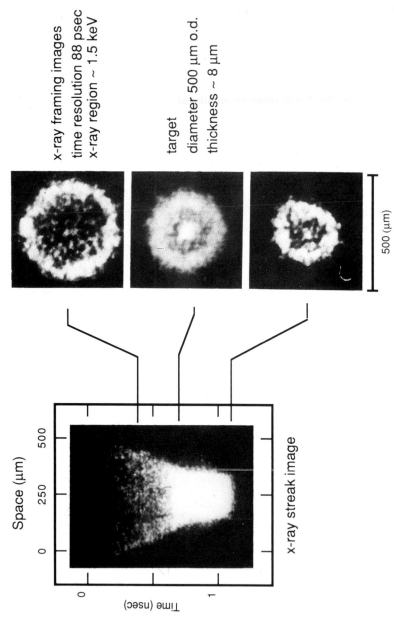

x-ray framing images
time resolution 88 psec
x-ray region ~ 1.5 keV

target
diameter 500 μm o.d.
thickness ~ 8 μm

500 (μm)

Space (μm)

0   250   500

Time (nsec)

0

1

x-ray streak image

Figure 7.7 The X-ray imaging technique of high temporal and spatial resolution in the implosion fusion.

demonstrated by using 6 μm diameter copper plasma to supply high energetic photons (>keV) (Miyanaga *et al.*, 1983).

To get sufficient time resolution, X-ray shadowgraphy in conjuction with X-ray streak and framing has been demonstrated. Advancement on micro electronics such as microchannel plate and sensitive X-ray cathode can enable us to afford the powerful X-ray photography in the streak mode as well as the framing mode. Figure 7.7 shows a implosion process measured by the X-ray imaging technique.

Suppression of self-emission effects can be made by increasing the spectral resolution of the observation system. An X-ray spectrometer coupled with imaging optics, such as a pinhole array is a useful method. This is shown in Figure 7.8. The potential of this method is not only to reduce the background but also to get $\langle \rho R \rangle$ and T_e values from a spectrally dependent transmittance of X-rays in a tracer gas seeded in the fuel (Azechi *et al.*, 1980).

Recent investigations indicate that plastic pusher targets and foam cryogenic targets have good compression performance. These targets emit less X-ray backgrounds and are more transparent to X-rays than conventional glass shell targets. It should be emphasized that X-ray spectroscopy and shadowgraphy will be potential methods in future experiments.

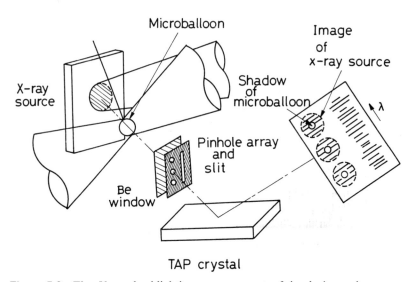

Figure 7.8 The X-ray backlighting measurement of implosion using x-ray spectrometer.

7.5   Burning process

Study of the time history of thermonuclear burn is a key issue for better understanding of implosion and ignition mechanisms. Since the burn duration is of the order of 100 ps, ultra fast neutron detectors are necessary to measure the burn history. Three methods have been experimentally evaluated. One is based on a neutron streak camera having a UO_2 cathode. Neutrons react with the uranium, to produce fission fragments which generate secondary electrons for streaking. (Brysk, 1973) The high fission cross-section 1.2 b of the uranium to 14.1 MeV neutrons and the high efficiency of the secondary electron emission are suitable for the neutron streak camera. To increase a sensitivity of the streak camera, use of a large cathod is proposed.

The second technique is based on a fast scintillator coupled to a streak camera or a phototube. (Niki, *et al.*, 1986) It can be used to measure the burn time. The attainable resolution is estimated to be 80 ps.

The third technique is to use a GaAs detector for neutron detection having a carrier life time of 60 ps. (Kania, 1988)

These techniques have been testified to show the sufficient time resolution and sensitivity to measure the absolute burn time and the burn duration in implosion experiments. (Yamanaka, *et al.*, 1986 a, Azechi, *et al.*, 1987, Prussin, *et al.*, 1987)

8 FUSION REACTOR

According to the recent development of the pellet implosion experiment, the basic knowledge on the inertial confinement fusion has been established. A reactor for electric power generation may be constructed with a sufficient pellet gain (Q: 50 ~ 200) using a laser driver of an output power of a few MW.

However, still several problems should be solved for realizing a power reactor. They are to develop a high efficient driver, a stable pellet supply system, blanket design for long life reactor, a neutron streaming and shielding device, a tritium handling system, and a power subtracting system for the ICF reactor. Brief overviews of the current status of the research on these problems are described.

8.1 Basic Requirements

(1) Overview of the system
The schematic diagram of an ICF reactor system is shown in Figure 8.1. This system consists of sub-systems such as a reactor chamber with reaction blankets, a laser driver, a heat transport system for electric power generation, a fuel recycle device and a pellet injector.

(2) Energy Balance of the total system
The energy flow in the reactor system is indicated in Figure 8.2. The system efficiency η_s of the laser fusion reactor is defined by the ratio

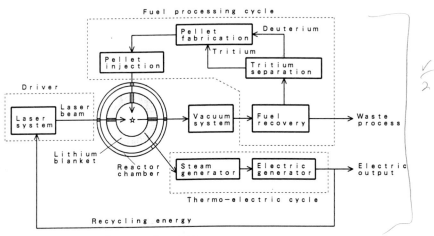

Figure 8.1 Block diagram of an ICF reactor.

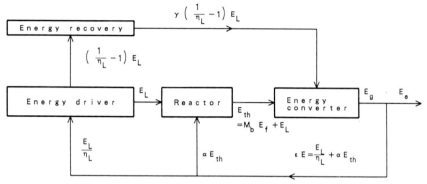

Figure 8.2   Power flow in an ICF reactor.

of the net electric output energy to the thermal output energy of the thermonuclear reaction which is given by equation (8.3). This system efficiency should be larger than 30% for avoiding the thermal pollution problem due to the wasted energy. A high system efficiency is also required for reducing the energy consumption of the reactor system itself.

The gross electric output power E_g is given by Ido *et al.* (1980)

$$E_g = \eta_e \left\{ \left(M_b Q + 1 \right) E_L + \gamma \left(\frac{1}{\eta_L} - 1 \right) E_L \right\}, \tag{8.1}$$

where Q is the pellet gain, η_e is the conversion efficiency of the thermal to the electric energy, M_b is the blanket gain, η_L is the laser efficiency, and γ is the recovery efficiency of waste thermal energy. The recycling energy for operating the laser driver and other sub-systems is assumed to be

$$\epsilon E_g = \frac{E_L}{\eta_L} + \alpha (M_b Q + 1) E_L, \tag{8.2}$$

where ϵ is the recirculating power fraction and α is the consumption energy of evacuation and cooling pumps of the reactor. Then the system efficiency η_s is given as

$$\eta_s = \frac{E_g (1 - \epsilon)}{M_b E_f}, \tag{8.3}$$

where E_f is the fusion output energy.

This system efficiency η_s has a functional dependence on the pellet gain Q for a given laser efficiency η_L. The calculated system efficiency η_s of a pure fusion system is shown in Figure 8.3 with fixed values of

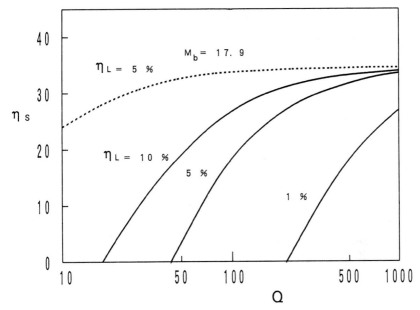

Figure 8.3 System efficiency η_S vs pellet gain Q for various values of the laser efficiency η_L.

$\eta_e = 40\%$, $\gamma = 0$, $\alpha = 5\%$, and $M_b = 1.2$ (solid line). It can be seen in this figure that a practical reactor plant can be designed for the pellet gain greater than 200 with a laser efficiency of 10%.

In a fission-fusion hybrid reactor, the blanket gain M_b is much higher than the case of pure fusion system. When natural uranium is used for the blanket, M_b is estimated to be 18 and the necessary pellet gain is about 20 for a low laser efficiency (5%) as shown in Figure 8.3 (broken line).

(3) Pellet gain and repetition rate

According to the above discussions, a certain value of the pellet gain Q is required to construct an actual reactor system, and this requirement provides necessary ρR and the laser energy E_L as described in Chapter 2. For example, a given laser efficiency η_L of 10% derives the laser output energy E_L of 1 ~ 5 MJ for a pure fusion reactor, while the pellet gain must be greater than 200.

The repetition rate of the driver or the reactor system is also an important parameter for designing a practical reactor. When the laser output energy and the pellet gain is given, the output power of the fusion reaction is directly proportional to the repetition rate of the system. The reactor chamber must withstand the blast of an explosion of E_L Q and must recover the firing conditions within the time of repetition.

8.2   Reactor Chamber

The output energy of fusion reaction is deposited in the reactor chamber in the form of radiation, shock waves, and reaction yield particles. In designing a chamber to contain the ICF plasma, the microexplosion of 200 ~ 1000 MJ energy at a repetition rate of greater than 1 Hz should be accepted in the structure of the vessel. The main source of this hazard is the radiation and neutrons. The first wall of the reactor chamber must withstand this loading and the activation of the wall should be suppressed.

8.2.1   Reaction Yield

From the ICF fusion plasma, four kinds of yield are emitted: fast neutrons, X-rays, charged particles, and pellet debris. Component ratios of the output energy of the fusion reaction depend on the target structure and ρR of the compressed fuel. In a case of $\rho R = 3 \, g/cm^2$, 69% of the fusion energy is deposited into neutrons (Ido *et al.*, 1982). Although these neutrons have a large mean free path in the wall, residuals are absorbed at the surface of the first wall and cause thermal stress and damage.

8.2.2   Wall Loading

Since the ICF reaction occurs in a very short time (less than 1 ns), the thermal loading of the first wall of the reactor chamber is impulsive and the heat will be deposited in a thin layer on the inside surface of the first wall. This sudden deposition of the heat provides a large amount of stress to the wall material. The first wall is exposed not only to the radiation but also to plasma products such as pellet debris and the blast wave. As a result of the short impulse and the thin layer of heat deposition on the wall surface, the damage of the wall material is a severe problem. If a dry wall of graphite or metal is exposed to the reactor plasma, excessive ablation (~ 1 cm/yr) and sputtering (~ 1 cm/yr) will take place (Duderstadt *et al.*, 1982).

Neutron damage of the reactor chamber material is also a severe problem in designing the cavity. The damage mechanisms are atomic displacement and gas production in the structural materials. An unprotected wall of stainless steel (316 SS) at a neutron loading of 1 MW/m² will receive the displacement damage of 10 dpa (displacements per atom) and the helium production rate of 220 appm (atomic parts per million) for a full power year (Duderstadt *et al.*, 1982). This damage rate is very critical for designing a long life reactor because the damage limits of 316 SS at 500°C are estimated to be 150 dpa and 500 appm helium.

8.2.3 First wall design

The key issues for designing a reactor chamber are to attain

(a) Adequate vacuum pressure in the cavity,
(b) Shielding for short range X-rays and plasma,
(c) Thermalization of 14 MeV neutron,
(d) Tritium breeding,
(e) Cooling scheme of the chamber,
(f) Removing system of the impurity and wastes.

Several kinds of reactor designs have been proposed with different types of first wall structure. These proposals can be divided into six categories as follows, (1) dry wall with liners, (2) magnetically protected wall, (3) gas-filled cavity, (4) wetted wall, (5) fluid curtain or jet, and (6) thick fluid flow. A brief discussion on these approaches is described.

(1) Dry Wall Cavity
(i) Large volume cavity
As described in 8.2.2., an unprotected dry wall will be damaged over an acceptable limit. Then the first wall should be protected from the excessive ablation and sputtering by means of placing a sacrificial liner such as graphite plate. This liner receives the radiation and debris from the plasma instead of the structural wall and will be replaced when the thickness is reduced to the designed value. The cavity size is also an important parameter for increasing the life of the liner. The design principally requires a large cavity size for reducing the wall load (Hovingh, 1976). Neutron protection is not provided in this scheme.

(ii) Magnetically protected wall
This scheme is based on the magnetic deflection of the charged particle. A cylindrical reactor chamber with external coils for producing an axial magnetic field is proposed. The charged particles from the reactor plasma are guided by the magnetic field to conical energy sinks located on both ends of the cylindrical cavity

(Devaney, 1974). Several disadvantages are seen in this scheme such as the magnetic field instabilities due to the work done by charged particles against the magnetic field. There is another difficulty with using liquid metal for the coolant because the magnetic field give forces to the liquid metal moving across the field.

(iii) Gas-filled cavity
A low pressure $(0.5 \sim 1$ torr$)$ of neon or xenon is contained in the reactor chamber as a buffer gas for moderating the first wall load due to the charged particle or target debris (Abdel-Khalik *et al.*, 1981). The gas pressure should be kept low enough to avoid the laser beam deflection.

(2) Wetted Wall Cavity
(i) Thin film liquid metal wall
This approach employs a thin layer of liquid metal (such as lithium) on the inner surface of the first wall for receiving the charged particles, X-rays and debris of the microexplosion to prevent the damage of the wall surface (Booth, 1973). In these designs, $1 \sim 2$ mm thick liquid lithium is supplied to the inner surface through a porous wall or by a thin film jet. The problems of this system are a complicated wall structure and the long recovery time of the lithium layer at the repetitive operation mode.

(ii) Fluid curtain or jet
The "waterfall" of liquid lithium separated from the chamber wall is used as the first wall to prevent the damage of the chamber due to the microexplosion (Maniscalco *et al.*, 1977). This lithium flow provides not only the wall protection but also the heat removal of the reactor chamber. The neutron damage of the reactor chamber itself is significantly reduced so that a long life operation of 30 years at wall loading of $4 \, \text{MW/m}^2$ is possible (Duderstadt *et al.*, 1982). The disadvantages of this scheme are a complicated structure of the chamber, and the large pumping power for the lithium flow. The vapor pressure of lithium in the chamber $(0.003 \, \text{Torr at } 500°\text{C})$ will be acceptable for the operation.

(iii) Thick fluid flow
A thick liquid lithium layer is guided by an external magnetic field to flow on the inner surface of the reactor chamber (Ido *et al.*, 1980).

The flow velocity and the thickness of the liquid lithium can also be controlled by the magnetic field. The details of a thick lithium flow reactor "SENRI-I" is given in § 8.4.

8.3 Other Supplemental Systems

(1) Target Injection
The pellet target of the ICF reactor should be injected into the reactor chamber for the fuel supply. It is also suitable for the repetitive operation of the reactor system at $1 \sim 10$ Hz. A precise target injection through a long distance (~ 10 m) is expected. Two different methods for driving the pellet are considered: electromagnetic and pneumatic injection. The tracking detection of the pellet is also important for achieving the reactor operation (Duderstadt *et al.*, 1982).

(2) Irradiation Optics
The scheme to introduce the laser beam into the reactor chamber is a key design because the beam port is opened on the chamber wall and intense radiation and a high flux of neutrons directly come out through this hole. Several approaches to prevent these difficulties have been proposed (Yamanaka *et al.*, 1983b; Ragheb *et al.*, 1978). A long distance (~ 100 m) and gas flow in front of the optics are possible way for protection. Dielectric coatings on the optics may provide serious problems due to the X-ray radiation (Duderstadt *et al.*, 1982). The neutron trapping at the bending mirror is also required to reduce the streaming neutron (Yamanaka *et al.*, 1983).

(3) Fuel Processing
The fuel pellet in a reactor system is supposed to be a cryogenic D-T shell target at low temperature. Manufacturing the shell and filling D-T fuel are important processes of the target fabrication. The tritium inventory of the reactor system is a vital problem for constructing an actual reactor system. Further investigation should be made on this problem.

8.4 An Example of Reactor Design

A magnetically guided lithium flow reactor (SENRI-I) has been designed at the Institute of Laser Engineering, Osaka University. The chamber structural wall is made of stainless steel with a radius of 6 m and the inner surface of the wall is covered with a thick lithium layer which flows from top to bottom of the chamber. An external magnetic field is applied to guide the lithium to flow along the wall and to control the thickness of the lithium layer. The total thickness of the inner and outer lithium layer is designed to be 70 cm to convert the neutron energy to thermal energy sufficiently and breed tritium. This thick flow of lithium ensures a low level of neutron fluence at the structural wall and the long life of the reactor cavity. The magnetic field is also useful for shielding charged particles such as α-particles from the lithium layer and reduce the expansion of the lithium layer caused by the heat pulse generated in the thin surface.

Although the lithium layer evaporates when the microexplosion irradiates the surface, the temperature of the lithium surface is cooled down quickly by the evaporation and the shock wave formation. This action provides a cryogenic pump effect in the reactor cavity.

Neutronics of a reactor system is an important subject of the design. A detailed research has been made on the SENRI-I system (Yamanaka et al., 1981c; Nakashima et al., 1985). The tritium breeding ratio in the blanket is ~ 1.6 and the neutron flux at the laser port is $6.7 \times 10^{12} n/cm^2$ sec. For the second shield of neutrons (biological shield), a 2.5 m thick concrete is enough. Induced activity of stainless steel (SUS-316) is a serious problem in such a reactor system. In this design, the dose rate of γ-radiation is 6.4 rem/hour between the first and second shield after 1 year in operation. The neutron streaming to the laser beam port is reduced with a zigzag path using reflecting mirrors and neutron shielding material enclosing the mirror. The reactor room of SENRI-I is shown in Figure 8.4.

A detailed study of thermo-electric cycle has been performed (Yamanaka et al., 1983b). The corrosion of stainless steel is a serious problem at high temperature (>500°C). Some refractory metal can prevent the corrosion. The electric power required for the lithium pump is estimated to be 3 ~ 27 MW for different values of flow velocity (6 ~ 55 ton/hour).

8.5   Future Development

Many conceptual design investigations have been made for long life reactors. However, few works have been done in the experimental research on the reactor development (Suzuki, 1981). The following items concerning the liquid metal blanket are important for the development of a laser fusion reactor.

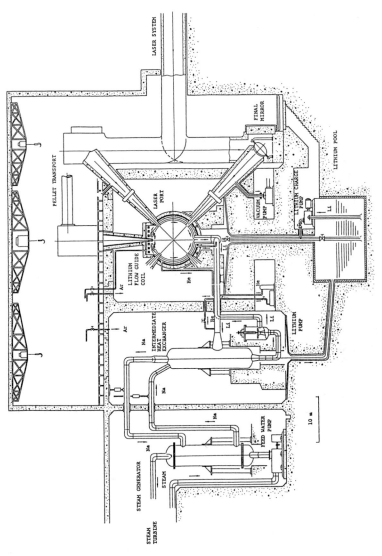

Figure 8.4 Cross-sectional view of the reactor room of SENRI-I.

8.5.1   Properties of Liquid Lithium Blanket

The basic physical properties of liquid lithium should be known for designing an actual reactor chamber.

The surface and volume heating due to the radiation of the micro-explosion provide intense pressure waves in the liquid lithium flow. The wall structure must withstand to this loading for a long period (~ 30 years). The computer simulation and experimental measurement are very important to get the database.

Since the lithium flow is guided by the external magnetic field, the turbulence in the flow due to the pressure wave may cause instability of the flow in the magnetic field of the reactor chamber.

The influences of blast wave and evaporation of the surface on the flow characteristics should be taken in to account for designing the blanket.

For reducing the shock wave loading to the cavity wall, the double phase flow may be useful in the actual reactor system. The interaction of pressure wave and double phase flow is studied.

The mist of liquid lithium is generated in the cavity due to the micro-explosion. This mist reaches to the beam port and fills the reactor chamber. The properties of generation and condensation of the lithium mist are very important to study.

The characteristics of instantaneous evaporation of the surface of liquid lithium by the heat pulse are also to be investigated.

8.5.2   Cavity Structure

The mechanical properties of the structural materials which compose the reactor chamber must be studied for developing the fusion reactor engineering.

The fatigue strength of the cavity material at high temperature and cyclic heating is an important property for the reactor design. The creep effect is also important.

The mechanical strength of the wall material under intense irradiation of repetitive radiation pulse must be known. Since the reactor cavity must endure the pressure pulses of the micro-explosion, analysis of the elastic response of the cavity should be made. The effect of impulse loading on fatigue characteristics of the cavity wall is important.

8.5.3   Engineering Estimation of ICF

Environmental requirements and economical scopes of the ICF power station are key issues for energy engineering in future.

(1) Energy resources

As described in Chapter 2 of this tract, the cumulative world demand for energy is estimated to be $11 Q$ ($= 10^{21}$ Joules) in the period of 1950 to 2000 (Gross, 1984). This energy consumption will increase to $61 Q$ for the period of 2000 to 2050. The fossil fuels of the world will provides only 30 to $50 Q$. The nuclear fission can supply $3 Q$ with uranium 235 and $200 Q$ with breeding of other fissionable fuel. The fusion resources are so abundant on the earth that an almost infinite amount of energy can supply the future for a long time.

(2) Environmental problems

It should be noted that fossil fuels lead to both air pollution and warming of the earth due to the increase of CO_2 gas. Fission reactors produce radioactive wastes which are difficult to dispose of in safety. Therefore, the development of fusion energy is of vital importance for the future, especially when environmental limitations are taken into account.

The only problem of the ICF reactor to the environment is the leakage of tritium from the system and the activation of the reactor vessel itself by neutron bombardments. Even though the present scheme of D-T reaction can provide a very acceptable and safe energy supply for mankind. In a long range of future, the neutron-free reactions such as $p - B$ and $D - {}^3He$ reactions are highly recommended.

(3) Economics of reactors

The MCF reactor can have an economical benefit only in an extremely large scale power output (much greater than 1000 MW). But the ICF reactor can be designed on a moderate scale (~ 1000 MW) which provides a great advantage for the daily operation and the energy economics.

The size of the ICF reactor must be the same scale as the fission reactor. However, the capital cost of the ICF reactor will be few times higher than the fission reactor considering the construction of a new system.

It is too early to discuss the economics of the ICF reactor in the present stage. Even though the abundance of fusion fuel resources and fundamentally safe reactor operation afford a very optimistic future for the ICF reactor which can supply almost infinite energy like the sun.

9 FUTURE PROSPECTS

The purpose of inertial confinement fusion research is to establish the fusion ignition by irradiating the D-T fuel pellet, to attain a fusion reactor for a heat source and to get an electric power generation plant. Several concepts for inertial fusion reactors are proposed, where the key issue for these is to obtain the high gain of the energy in the DT fuel. The configuration of a fusion reactor by the inertial confinement fusion is estimated to be much easier than that of the magnetic confinement fusion. Scientific several successes have been attained in the last decades. However there are still technical problems to be solved especially in the development of efficient drivers and economic fuel pellets. Table 9.1 shows the requirements on implosion fusion plasma parameters.

The following are important issues of the ICF to be explored in the near future.

(1) Decrease of the necessary driver energy
Nuclear fusion had already been attained by the development of hydrogen bombs. The objectives of the ICF are to attain the controlled microfusion reaction which is applicable to the "new internal combustion engine". The most important issue is to decrease the necessary driver energy as much as possible. If it needs to be large, the ICF will lose the economical merit as well as the security of the fusion implosion. To date the necessary energy of drivers is estimated to be as large as $1 \sim 10\,\text{MJ}$. The

Table 9.1 Requirements on implosion plasma parameters.

Implosion parameters	Required value
η	$\geq 5\,\%$ (coupling)
ε_c	$< 10\,\text{eV}$ (non preheating) at Liquid Density
R_h/R_o	$> \dfrac{1}{30}$ (compression)
ρ	$\geq 200\ \text{g}/\text{cm}^3$ (fuel density)
T_h	$\simeq 10\,\text{keV}$ (ignition)
R_h/R_α	≥ 1 (α range)

Table 9.2   Candidates for reactor laser.

Lasers	Wavelength (μm)	Energy (MJ)	Pulse Width (ns)	Efficiency (%)	Remarks
Nd Glass	1.05 0.35	1	>10 ns	1~2	Design for efficient extraction using convensional technology
CO_2	10.6	2	20 ns	10~12	High extraction efficiency with multi-line, multi-pass amplification, too long wavelength
KrF	0.25	1	20 ns	5	Simple pulse Compression technique
Novel Solid State	<1	1	>10 ns	10~12	Search for new materials for high repetition operation and new excitation light sources for efficient operation
Free Electron	variable	–	10 ns	20	Developing

construction of this size of laser is not difficult in the present art of technology. However the cost of lasers per joule is expected to decrease one order or more for the actual performance. Otherwise the economical feasibility becomes weak. In Table 9.2 several lasers for the candidate reactor driver are shown.

(2) Attainment of uniform compression

The goal of implosion physics is to compress the DT fuel up to 1000 times the liquid density, that is a density of $200 \, g/cm^3$. To date we have reached 600 times the liquid density. The key point of the compression is the uniformity of the laser irradiation as well as the fuel pellet. The fusion energy yield is expected to be $100 \sim 1000$ which depends on the efficiency of drivers. The advancement in physics is really remarkable in the last few years.

(3) Fabrication of fuel target

The theoretical demands for fuel pellets are very stringent to fabricate. High uniformity of the pellet, multilayered precise structure of the shell and anti-preheating configuration of the fuel target are highly expected.

The target fabrication is one of the most important breakthrough items for the inertial confinement fusion. The cost of the pellets also should be kept low; hence, cryogenic fuel or plastic shell pellets are being developed.

(4) Direct drive and indirect drive

Needless to say, the direct drive is much more efficient than the indirect drive. However the direct drive demands high uniformity of the

Table 9.3 Development scenario to ICF reactor.

	I Ignition Exp	II LFCX	III LFER	IV LFPR (D to C)
Mission	Ignition (Scientific feasibility) Close to accomplishment	Burning (Engineering feasibility) · High gain Pellet design · Reactor engineering · Intense neutron source	Reactor engineering test	Demonstration of power plant
Laser	Nd-glass 100 kJ Single shot	Solid, KrF 1MJ Single − 1 shot/min.	1MJ 1Hz	10 MJ 1 Hz
Pellet	$Q = 1$ 10^{16} N/shot $\rho \geqq 1000\rho_0$ (200g/cm^3) $\rho R \geqq 0.3$ g/cm^2	$Q = 10 \sim 100$ $10^{18} \sim 10^{19}$ N/shot $5 \sim 50$ MJ/shot $\rho \geqq 2000\,\rho_0$ (400g/cm^3) $\rho R \geqq 1$ g/cm^2	$Q = 10 \sim 100$ $10^{18} \sim 10^{10}$ N/sec $5 \sim 50$ MW th	$Q = 50 \sim 500$ $10^2 \sim 10^3$ MWE $\rho \sim 2000 \sim 4000\,\rho_0$ (400~800 g/cm^3) $\rho R \sim 3 \sim 5$ g/cm^3

irradiation and control of the instability in the implosion. For these purposes, the random phase plate, the induced spatial incoherency by grating, the beam smoothing by spectral dispersion and the multi-lens or Fresnel lens system are used to improve the uniformity of laser beams. The ILE Osaka has recorded the highest neutron yield 10^{13} with the direct driven LHART (Large High Aspect Ratio Target) by the stagnation free compression using the random phasing irradiation.

The indirect drive scheme is represented by the cannonball target which is imploded by the X-rays produced in the target cavity. This scheme is superior in the uniformity of implosion to the direct drive. However the coupling of the beam is less efficient which needs the stronger laser drive than the direct drive. The experiment is now directed to find a scaling law of indirect drive implosion. The spin off to the X-ray lasers becomes very prosperous and interesting. The appearances of powerful drivers is highly expected for the indirect drive fusion research.

(5) Computer simulation
To simulate the very complicated implosion processes, the development of the codes adopting the multidimensional transport of radiation and heat is essentially important. Several approaches in simulation calculations using super-computers are performed by the methods of hydro-dynamics in high temperature and high density plasmas concerning the implosion, ignition and burn.

This field has recently made great progress in software as well as hardware. The predictions due to computer simulation are very powerful tools to perform the actual implosion experiments. These two researches should be closely related to pursue the development of ICF.

(6) Diagnostics

ICF research requires high technology to obtain measurements which have never been made before. The highly resolved techniques for spatial and temporal measurement and streaking and framing cameras for the X-rays and the neutrons are being developed. It is not unrealistic to say that without the newly developed diagnostics the experimental work at ICF could not be done.

As described in the preceding chapters, there are still several physics issues to be solved in the ICF. However with recent advances in research we can have confidence that the fundamental problems will definitely be resolved in the coming decade and the feasibility of ICF will be established by the next century. Figure 9.1 shows the steady advancement of laser fusion towards the ICF reactor regime.

As for the scenario for the ICF reactor, Table 9.3 indicates the four steps to develop. Experiments for phase I have been promoted so much that we can almost say that the requirements are completed. The attention to the next step is booming which needs the larger laser facility with scope for burning experiments. The international concern for this issue is to build an MJ level driver aiming toward the ICF engineering feasibility test. By international collaboration this milestone will hopefully be overcome by the next century.

References

1. Abdel-Khalik, S.I., Moses, G.A. and Peterson, R.R. (1981), *Inertial confinement fusion reactors based on the gas protection concept*. Nucl. Eng. and Design, **63**, 315.
2. Ahlborn, B., Key, M.H., and Bell, A.R., (1982), *An analytic model for laser-driven ablative implosion of spherical shell targets*, Phys. Fluids, **25**, 541.
3. Azechi, H., Oda, S., Hamano, M., Sasaki, T., Yamanaka, T., and Yamanaka, C., (1980), *Multifrequency X-ray backlighting of laser-imploded targets*. Appl. Phys. Lett. **37**, 998.
4. Azechi, H., Miyanaga, N., Stapf, R.O., Itoga, K., Nakaishi, H., Yamanaka, M., Shiraga, H., Tsuji, R., Ido, S., Nishihara, K., Izawa, Y., Yamanaka, T., and Yamanaka, C., (1986), *Experimental determination of fuel density radius product of inertial confinement fusion targets using secondary nuclear fusion reactions*. Appl. Phys. Lett. **49**, 555.
5. Azechi, H., Miyanaga, N., Stapf, R.O., Yamanaka, M., Yamanaka, C.,

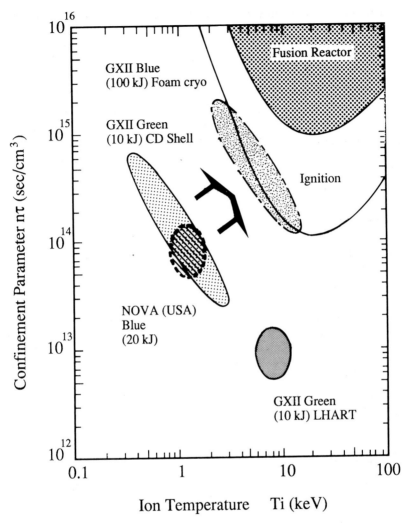

Figure 9.1 Progress of laser fusion towards ICF reactor regime.

and Iguchi, T., (1987), *Fast neutron detectors for thermonuclear burn time and duration measurements in inertial confinement fusion targets. in*: Proc. International Symposium on Short Wavelength Lasers and Their Applications, Osaka, Japan, 11–13 November, 1987.

6. Basov, N.G. and Krokhin, O.H., (1964), *Conditions for heating up of a plasma by the radiation from an optical generator.* JETP, **19**, 123.

7. Basov, N.G., Vygovskii, O.B., Gus'kov, S. Yu., Il'in, D.V., Lekovskii, A.A., Rozanov, V.B., and Sherman, V.E., (1986), *Diagnostics of laser-*

fusion plasmas on the basis of the products of secondary fusion reactions.
Sov. J. Plasma Phys. **12**, 526.

8. Bettinger, A., Charles, C., Osmalin, J. and Giraud, J.G. (1976), *Laser beam brightness improvement with high power spatial filtering.* Opt. Commun., **18**, 176.

9. Bell, A.R., Evans, R.G., and Nicholas, D.J., (1981), *Electron energy transport in steep temperature gradients in laser-produced plasmas.* Phys. Rev. Lett. **46**, 243.

10. Bobin, J.L., (1971), *Flame propagation and overdense heating in a laser created plasma.* Phys, Fluids, **14**, 2341.

11. Bodner, S.E., (1981), *Critical elements of high gain laser fusion*, J. Fusion Energy, **1**, 221.

12. Booth, L.B. (1973), *Central station power generation by laser driven fusion.* Nucl. Eng. and Design, **24**, 263.

13. Brysk, H., (1973), *Fusion neutron energies and spectra.* Plasma Phys. **15**, 611.

14. Cable, M.D., Lane, S.M., Glendinning, S.G., Lerche, R.A., Singh, M.S., Munro, D.H., Hatchett, S.P., Estabrook, K.G., and Suter, L.J., (1986), *Implosion experiments at NOVA.* Bull. Am. Phys. Soc. **31**, 1461.

15. Campbell, J. (1986), *Eliminating platinum inclusion in laser glass.* Energy and Technology Review, LLNL UCRL-52000-86-4/5.

16. Carlson, R.L., Carpenter, J.P., Casperson, D.E., Gibson, R.B., Godwin, R.P., Haglund Jr. R.F., *et al.* (1981), *HELIOS: a 15 TW carbon dioxide laser-fusion facility.* IEEE J. of Quant Elect., **QE-17**, 1662.

17. Ceglio, N.M., and Coleman, L.W., (1977), *Spectrally resolved α emission from laser fusion targets.* Phys. Rev. Lett., **39**, 20.

18. Chen, Y.W., Miyanaga, N., Yamanaka, M., Izawa, Y., Yamanaka, C., and Tamura, S., (1986), *X-ray URA-CT applied to the laser fusion research in*: Proc. 17th Joint Conf. Image Technology, Tokyo, (1986), ed. Yasuda, Y. (Executive Committee of 17th Joint Conf. Image Technology, Tokyo) 18–2.

19. Cook, E.G. and Birx, D.L. (1983), *The advanced test accelerator: a high-current induction linac.* IEEE Trans., **NS-30**, 1381.

20. Craxton, R.S. (1981), *High efficiency frequency tripling scheme for high power Nd: glass lasers.* IEEE J. of Qunt. Elect., **QE-17**, 1771.

21. Cuderman, J.F., and Gilbert, K.M., (1972), *An x-ray spectrometer for laser-induced plasmas.* Rev. Sci. Instrum., **46**, 53.

22. Deacon, D., Elias, L.R., Madey, J.M.J., Ramian, G.J., Schwettman, H.A. and Smith, T.I. (1977), *First operation of a free-electron laser.* Phys. Rev. Lett., **38**, 892.

23. Devaney, J.J. (1974), *Magnetically protected first wall for a laser induced thermonuclear reaction.* Los Alamos Scientific Laboratory Report LA-5699-MS.

24. Den, X., Liang, X., Chen, Z., Yu, W. and Ma, R. (1986), *Uniform illumination of large targets using a lens array.* Applied Optics, **25**, 377.

25. Duderstadt, J.J. and Moses, G.A. (1982), *Inertial Confinement Fusion.* (John Wiley and Sons).

26. Duston, D., Clark, R.W., Davis, J., and Apruzese, J.P., (1983), *Radiation energetics of a laser-produced plasma.* Phys. Rev., **A27**, 1441.

27. Eidman, K., Sachsenmaier, P., Salzman, H. and Sigel. R. (1972), *Optical*

isolators for high power giant pulse lasers. J. of Phys. E, **5**, 56.

28. Emery, M.H., Gardner, J.H., and Boris, J.P., (1982), *Nonlinear aspects of hydrodynamic instabilities in laser ablation.* Appl. Phys. Lett., **41**, 808.

29. Emery, M.H., Gardner, J.P., and Bodner, S.E., (1986), *Strongly inhibited Rayleigh-Taylor growth with 0.25 μm lasers.* Phys. Rev. Lett., **57**, 703.

30. Emmett, J.L., Krupke, W.F. and Davis, J.I. (1984 a), *Laser R & D at the Lawrence Livermore National Laboratory for fusion and isotope separation applications.* IEEE J. of Quant. Elect., **QE-20**, 591.

31. Emmett, J.L., Krupke, W.F. and Sooy, W.R. (1984 b), *The potential of high-everage-power solid state laser.* Lawrence Livermore National Laboratory Report, UCRL-53571

32. Estabrook, K.G., Valeo, E.J., and Kruer, W.L., (1975), *Two-dimensional relativistic simulations of resonance absorption.* Phys. Fluids, **18**, 1151.

33. Estabrook, K., and Kruer, W.L., (1978), *Properties of resonantly heated electron distributions.* Phys. Rev. Lett., **40**, 42.

34. Fabbro, R., Max, C., Fabre, E., (1985), *Planar laser-driven ablation: effect of inhibited electron thermal conduction.* Phys. Fluids, **28**, 1463.

35. Faltens, A., Hoyer, E., Keefe, D. and Laslett, L.J. (1979), *Design/cost study of an induction linac for heavy ions for pellet-fusion.* IEEE Trans. Nuc. Sci., **NS-26**, 3106.

36. Fessenden, T.J., Celata, C.M., Faltens, A., Herderson, T., Judd, D.L., Keefe, D., *et al.* (1987), *Preliminary design of a ~ 10MV ion accelerator for HIF research.* Laser and Particle Beams, **5**, 457.

37. Forslund, D.W., Kindel, J.M., and Lee, K., (1977), *Theory of hot-electron spectra at high laser intensity.* Phys. Rev. Lett., **39**, 284.

38. Fraley, G.S., Linnerberg, E.J., Mason, R.J., and Morse, R.L., (1974), *Thermonuclear burn characteristics of compressed deuterium-tritium microspheres.* Phys. Fluids, **17**, 474.

39. Gabriel, A.H., (1972), *Dielectronic satellite spectra for Highly-charged helium-like ion lines.* Mon. Not. R. Astr. Soc. **160**, 99.

40. Gamalii, E.G., Gus'kuv, S. Yu., Krokhin, O.N., and Rozanov, V.B., (1975), *Possibility of determining the characteristics of laser plasma by measuring the neutrons of the DT reaction.* JETP Lett. **21**, 70.

41. Gelinas, R.J. edited (1987), *High-average-power lasers. Laser Program Annual Report 1986,* LLNL UCRL-50021-86.

42. Gintzburg, V.L., (1970), *The propagation of electromagnetic wave in plasmas,* 2nd ed. (Pergamon Press)

43. Goldstein, S.A., Cooperstein, G., Lee, R., Mosher, D. and Stephanakis, S.J. (1978), *Focusing of intense ion beams from pinched-beam diodes.* Phys. Rev. Lett., **40**, 1504.

44. Gross, R.A., (1984), *Fusion energy. Chap. 1* (John Wiley and sons)

45. Haddlestone, R.H., and Leonard, S.L., (1965), *Plasma Diagnostic Technique, Chap. 8,* (Academic Press)

46. Hasegawa, A., Nishihara, K., Daido, H., Fujita, M., Ishizaki, R., Miki, F., *et al.* (1988), *Magnetically insulated and inertially confined fusion-MICF.* Nucl. Fusion, **28**, 369.

47. Hattori, F., Takabe, H., and Mima, K., (1986), *Rayleigh-Taylor instability in a spherically stagnating system.* Phys. Fluids, **29**, 1719.

48. Hovingh, J. (1976), *First wall studies of a laser fusion hybrid reactor design.* Lawrence Livermore Laboratory Report, UCRL-78090.

49. Hunt, J.T., Glass, J.A., Simmons, W.W., and Renard, P.A., (1978), *Suppression of self-focusing through low-pass spatial filtering and relay imaging.* Appl. Opt., **17**, 2053.

50. Hunt, J.T., Speck, D.R. and Warren, W.E. (1987), *ATHENA design. Laser Program Annual Report 1986*, LLNL UCRL-50021-86, 6–81.

51. Hunt, J.T. and Speck, D.R. (1989), *Present and future performance of the NOVA laser system.* Optical Engineering, **28**, 461.

52. Hunter, A.H., Hunter, R.O. and Johnson, T.H. (1986), *Scaling of KrF lasers for inertial confinement fusion.* IEEE J. of Quant. Elect., **QE-22**, 386.

53. Ido, S., Imasaki, K., Izawa, K., Kato, Y., Kitagawa, Y., Matoba, M. *et al.* (1980), *Laser fusion reactor concept of high pellet gain using magnetically guided Li flow.* 8th International Conference on Plasma Physics and Controlled Nuclear Fusion Research, Brussels.

54. Ido, S., Nakai, S., Yamanaka, C., Hattori, H., Kodaira, H., Kondo, S. *et al.* (1982), *Neutronics calculation in pellets and blankets of laser fusion reactor concept SENRI-I.* J. Nucl. Sci. Technolo. **19**, 1019.

55. Imasaki, K., Miyamoto, S., Ozaki, T., Fujita, H., Yugami, N., Higaki, S. *et al.* (1984), *Light ion fusion research in Japan.* Proc. of 10th IAEA Int. Conf. on Plasma Phys. and Controlled Nucl. Fusion, London.

56. Imasaki, K., Miyamoto, S., Yamanaka, T., Mima, K., Kuruma, S., Nakai, S. *et al.* (1989), *Studies about application of free electron laser to inertial confinement fusion.* Proc. of 11th IEEE Int. Conf. on Free Electron Lasers, Florida.

57. Johnson, D.J., (1972), *An x-ray spectral measurement system for nanosecond plasmas.* Rev. Sci. Instrum., **45**, 191.

58. Johnson, D.J., Kuswa, G.W., Farnsworth, A.V., Quintenz, J.P., Leeper, R.J., Burns, E.J.T., *et al.* (1979), *Production of 0.5-TW proton pulses with a spherical focusing, magnetically insulated diode.* Phys. Rev. Lett., **42**, 610.

59. Johnstone, T.W. and Dawson, J.M., (1973), *Correct values for high-frequency power absorption by inverse bremsstrahlung in plasma.* Phys. Fluids, **16**, 722.

60. Kacenjar, S., Goldman, L.M., Enterberg, A., and Skupsky, S., (1984), $\langle \rho R \rangle$ *measurements in laser-produced implosions using elastically scattered ions.* J. Appl. Phys. **56**, 2027.

61. Kanabe, T., Nakatsuka, M., Kato, Y. and Yamanaka, C. (1986), *Coherent stacking of frequency-chirped pulses for stable generation of controlled pulse shapes.* Opt. Commun., **58**, 237.

62. Kanabe, T., Jitsuno, T., Nakatsuka, M., Yamanaka, C. and Nakai, S. (1989), $10^{13} W/cm^2$ *focusability of a high average power glass laser.* Proc. of SPIE Vol. 1040; High Power and Solid State Laser II, 177.

63. Kania, D.R., and Lane, S.M., and Prussin, S.G., (1988), *Measurement of single 14 MeV neutron bursts with 100 ps time resolution.* Appl. Phys. Lett., **53**, 1988.

64. Kato, Y., Mima, K., Miyanaga, N., Arinaga, S., Kitagawa, Y., Nakatsuka, M. and Yamanaka, C. (1984), *Random phasing of high-power lasers for uniform target acceleration and plasma-instability suppression.* Phys. Rev. Lett., **53**, 1057.

65. Kato, Y., Nishihara, K., Miyanaga, N., Azechi, H., Tanaka, K.A.,

Yamanaka, C., *et al.*, (1985), *Implosion of Cannonball targets at 1.053 μm and 0.526 μm with GEKKO XII glass laser system.* ILE Quarterly Progress Report, No. 14, pp 2–25.

66. Kelly, J.H., Shoup, M.J. and Smith, D.L. (1989), *OMEGA upgrade staging options.* Proc. of SPIE Vol. 1040; High power and Solid State Lasers II, 184.

67. Key, M.H., Lewis, C.L.S., Lunney, J.G., More, A., Hall T.A., and Evans, R.G., (1978), *Pulsed x-ray shadowgraphy of dense, cool, laser-imploded plasma.* Phys. Rev. Lett., *41*, 1467.

68. Kishimoto, Y., and Mima, K., (1983), Electronic state and x-ray spectrum of high density laser produced plasma. Kakuyugo Kenkyu *50*, separate volume, 1–84. (in Japanese)

69. Kruer, W.L., (1988), *The Physics of Laser Plasma Interactions, pp. 45*, Addison-Wesley Publishing, *Redwood City.*

70. Krupke, W.F. (1974), *Induced-emission cross sections in neodymium laser glasses.* IEEE J. of Quant. Elect., **QE-10**, 450.

71. Lane, S.M., Campbell, E.M., and Bennett, C., (1980), *Measurement of DT neutron-induced activity in glass-microshell laser fusion targets.* Appl. Phys. Lett., **37**, 600.

72. Lehmberg, R.H., Schmitt, A.J. and Bodner, S.E. (1987), *Theory of induced spatial incoherence.* J. of Appl. Phys., **62**, 2680.

73. Lerche, R.A., Coleman, L.W., Houghton, J.W., Speck, D.R., and Storm, E.K., (1977), *Laser fusion ion temperatures determined by neutron time-of-flight techniques.* Appl. Phys. Lett., **31**, 645.

74. Lokke, W.A., and Grasberger, (1977), Res. Rep., UCRL-52276, LLNL.

75. Manheimer, W.M., (1977), *Energy flux limitation by ion acoustic turbulence in laser fusion schemes.* Phys. Fluids, **20**, 265.

76. Maniscalco, J.A., Meier, W.R. and Monsler, M.J. (1977), *Design studies of a laser fusion power plant.* Lawrence Livermore Laboratory Report, UCRL-80071.

77. Max, C.E., (1982), *in Laser Plasma Interaction, Balian, R. and Adam, J.C.,* pp. 388. (Les Houches, Section X X X IV, North Holland, Pub.).

78. Maxon, S., (1972), *Bremmstrahlung rate and spectra from a hot gas (Z = 1).* Phys. Rev. **45**, 1630.

79. McCrory, R.L., Montierth, L., Morse, R.L., and Verdon, C.P., (1981), *Nonlinear evolution of ablation-driven Rayleigh Taylor instability.* Phys. Rev. Lett., **46**, 336.

80. Mihalas, D., and Mihalas, B.W., (1984), *Foundation of Radiation Hydrodynamics* (Oxford Univ. Press, Oxford).

81. Mima, K., (1988), *High density compression of hollow shell target by green and blue GEKKO XII laser.* Bull. American Phys. Soc., **33**, 1880.

82. Miyanaga, N., Kato, Y., and Yamanaka, C., (1983), *Point-source x-ray backlighting for high-density plasma diagnostics.* Appl. Phys. Lett., **42**, 160.

83. Mochizuki, T., Yabe, T., Okada, K., Hamada, M., Ikeda, N., and Yamanaka, C., (1986), *Atomic-number dependence of soft-x-ray emission from various targets irradiated by a 0.53-μm-wavelength laser.* Phys. Rev. A, **33**, 525.

84. Mochizuki, T., Yabe, T., Tanaka, K.A., Yamanaka, C., Sigel, R.,

Tsakiris, G.D., *et al.*, (1987), *X-ray confinement in a laser heated cavity*, Nuclear Fusion Suppl. **3**, 25.

85. More, R.M., (1981), *Atomic Physics in Inertial Confinement Fusion*, Res. Rep., UCRL-84991, LLNL.

86. Müller, R.W. (1988), *RF linac driver for commercial heavy ion beam fusion*. Proc. of 3rd Inertial Confinement Fusion System and Application Colloquium, Madison.

87. Nakashima, H., Kanda, Y. and Ido, S. (1985), *Neutronic optimization for a blanket of the laser fusion reactor SENRI-I*. J. of Fusion Energy, **4**, 315.

88. Niki, H., Itoga, K., Miyanaga, N., Yamanaka, M., Yamanaka, T., Yamanaka, C., Iida, T., Takahashi, A., Sumita, K., Kinoshita, K., Takiguchi, Y., Hayashi, I., and Oba, K., (1986), *Characteristics of uranium-oxide cathode for neutron streak camera*. Rev. Sci., Instrum., **57**, 1743.

89. Nuckolls, J., Wood, L., Thiessen, A. and Zimmerman, G. (1972), *Laser compression of matter to super-high densities: Thermonuclear applications*. Nature, **239**, 139.

90. Nugent, K.A., and Luther-Davies, B., (1984), *Penumbral imaging of high energy x-rays from laser-produced plasma*. Opt. Commun. **49**, 393.

91. Owadano, Y., Okuda, I., Matsumoto, Y., Tanimoto, M., Tomei, T., Koyama, K., *et al.* (1989), *Recent progress in ASHURA, a high power KrF laser system*. CLEO'89 Technical Digest, MG-3.

92. Presenyakav, L.P., (1976), *X-ray spectroscopy of high-temperature plasma*. Sov. Phys. USP **19**, 387.

93. Prussin, S.G., Lane, S.M., Richardson, M.C., and Noyes, S.G., (1986), *Debris collection from implosion of microballoons*. Rev. Sci. Instrum. **57**, 1734.

94. Prussin, S., Kania, D.R., Lane, S., and Jones, B., (1987), *High speed detection of 14 MeV neutrons: burnwidth and bangtime* Bull. Am. Phys. Soc. **32**, 1874.

95. Ragheb, M.M.H., Klein, A.C. and Maynard, C.W. (1978), *Three dimensional neutronics analysis of the mirror-beam-duct-shield system for a laser driven power reactor*. Univ. of Wisconsin Fusion Engineering Program Report UWFDM-239.

96. Ramirez, J.J., Prestwich, K.R., Burgess, E.L., Furaus, J.P., Hamil, R.A., Johnson, D.L. *et al.* (1987), *The HERMES III program*. Proc. of 6th IEEE Pulsed Power Conference, Arlington, 294.

97. Randal, C.J., Thomson, J.J., and Estabrook, K.G., (1979), *Enhancement of stimulated brillouin scattering due to reflection of light from plasma critical surface*. Phys. Rev. Lett., **43**, 924.

98. Rosocha, L.A., Anderson, R.G., Czuchlewski, S.J., Hanlon, J.A., Jones, J.E., Jones, R.G., *et al.* (1989), *Kilojoule operation of the AURORA KrF ICF laser system at Los Alamos*. CLEO'89 Technical Digest, MG-2.

99. Scifres, D.R., Burnham, R.D. and Streifer, W. (1982), *High power coupled multiple stripe well injection lasers*. Appl. Phys. Lett., **41**, 118

100. Shaw, M.J., O'Neill, F., Edwards, C.B., Nicholas, D.J. and Craddock, D. (1982), *SPRITE-a 250 J KrF laser*. Appl. Phys., **B-28**, 127.

101. Skupsky, S., (1980), *X-ray line shift as a high-density diagnostic for laser-imploded plasmas*. Phys. Rev. **A21**, 1316.

102. Skupsky, S. and Kessler, T. (1987), *A source of hot spot in frequency-tripled laser light*. *LLE Review*, **31**, 106.
103. Speck, D.R. and Bliss, E.S. (1986), *NOVA laser activation and performance*. Laser Program Annual Report, LLNL 1985, UCRL-50021-85, 5-2.
104. Spitzer, L., and Härm, R., (1953), *Transport phenomena in a completely ionized gas*. Phys. Rev. **89**, 977.
105. Stokowski, S.E., Saroyan, R.A. and Weber, M.J. (1978), *Nd-doped laser glass spectroscopic and physical properties*. Lawrence Livermore Laboratory, Misc. Report MS-095.
106. Suda, A., Obara, M. and Fujioka, T. (1984), *Atmospheric pressure operation of an electron beam excited KrF laser using Kr/F_2 mixtures*. Appl. Phys. Lett., **45**, 1165.
107. Suzuki, K. (1981) *Development of liquid lithium test loop*. IHI Engineering Review, 14, No. 4
108. Takabe, H., Nishihara, K., and Taniuti, T., (1978), *Deflagration waves in laser compression I*. J. Phys. Jpn., **45**, 2001.
109. Takabe, H., and Mulser, P., (1982 a), *Self-consistent treatment of resonance absorption in a streaming plasma*. Phys. Fluids, **25**, 2304.
110. Takabe, H., Mima, K., and Yabe, T., (1982 b), *Electrostatic field generation and hot electron reduction in a laser produced plasma*. J. Phys. Soc. Jpn. **51**, 2293.
111. Takabe, H., Montierth, L., and Morse, R.L., (1983), *Self-consistent eigenvalue analysis of Rayleigh-Taylor instability in an ablating plasma*. Phys. Fluids, **26**, 2299.
112. Takabe, H., Mima, K., Motierth, L., and Morse, R.L., (1985), *Self-consistent growth rate at the Rayleigh-Taylor instability in an ablatively accelerating plasma*. Phys. Fluids, **28**, 3676.
113. Takabe, H., and Mima, K., (1987), *Numerical study of ignition by stagnation-free implosion*. ILE-Report, 8713, ILE, Osaka Univ.
114. Turman, B.N., Martin, T.H., Nean, E.L., Humphreys, D.R., Cook, D.L., Goldstein, S.A. *et al.* (1985), *PBFA II, a 100 TW pulsed power driver for the inertial confinement fusion program*. Proc. of 5th IEEE Pulsed Power Conference, Arlington, 155.
115. Ueda, K. and Takuma, H. (1988), *Scalability of high power KrF lasers for ICF driver*. *Short-Wavelength Laser and Their Applications*, edited by Yamanaka, C., Springer Proceeding in Physics, 30.
116. Wang, C.L., Kalibjian, R., and Singh, M.S., (1982), S.P.I.E. **384**, 278.
117. Wilson, H.L., (1972), *Progress in High Temperature Physics and Chemistry*, **5**, (Pergamon Press, Oxford).
118. Witte, K.J., Fill, E., Brederlow, G. and Volk, R. (1981), *Advanced iodine laser concepts*. IEEE J. of Quant. Elect., **QE-17**, 1809.
119. Yamada, A., Niki, H., Kisoda, A., Yamanaka, T., Yokoyama, M., and Yamanaka, C., (1983), *URA coaded aperture camera*. Report of Electronics and Communications Society of Japan, ED83-27.
120. Yamamoto, K., and Narumi, H., (1983), *High-density diagnostics for laser imploded plasmas*. J. Phys. Soc. Jpn. **52**, 520.
121. Yamanaka, C., Yamanaka, T., Sasaki, T., Yoshida, K. and Waki, M. and Kang, H.B., (1972), *Anomalous heating of a plasma by a laser*. Phys. Rev., **A-6**, 2335.

122. Yamanaka, C., Kato, Y., Izawa, Y., Yoshida, K., Yamanaka, T., Sasaki, T. *et al.* (1981 a), *Nd-doped phosphate glass laser systems for laser fusion research.* IEEE J. of Quant Elect., **QE-17**, 1639

123. Yamanaka, C., Nakai, S., Matoba, M., Fujita, H., Kawamura, Y., Daido, H., *et al.* (1981 b), *The LEKKO VIII CO_2 gas laser system.* IEEE J. of Quant. Elect., **QE-17**, 1678

124. Yamanaka, C., Nakai, S., Ido, S., Yabe, T., Imasaki, K., Matoba, M. *et al.* (1981 c), *Conceptual design of 1200 MWth laser fusion reactor in ILE Osaka University.* Laser Interaction and Related Plasma Phenomena, edited by Schwarz, H.J. *et al.*, Plenum Publishing Corp, **5**, 361.

125. Yamanaka, C., Ido, S., Nakai, S., Norimatsu, T., Mima, K., Matoba, M. *et al.* (1983 b), *Concept and design of ICF reactor "SENRI-I".* Fusin Reactor Design and Technology, Vol. 1, IAEA-TC-392/28

126. Yamanaka, C., Nakatsuka, M., Nishimura, H., (1986 a), *Survey of the laser fusion.* Rev. Laser Eng., **14**, 1003. (in Japanese)

127. Yamanaka, C., Nakai, S., Yabe, T., Nishihara, K., Uchida, S., Izawa, Y., Norimatsu, T., Miyanaga, N., Azechi, H., Nakai, M., Takabe, H., Jitsuno, J., Mima, K., Nakatsuka, M., Sasaki, T., Yamanaka, M., Kato, Y., Mochizuki, T., Kitagawa, Y., Yamanaka, T., and Yoshida, K., (1986 b), *Laser implosion of high-aspect-ratio targets produces thermonuclear neutron yields exceeding 10^{12} by use of shock multiplexing.* Phys. Rev. Lett. **56**, 1575.

128. Yamanaka, C. and Nakai, S. (1986 c), *Thermonuclear neutron yield of 10^{12} achieved with GEKKO XII green laser.* Nature, **319**, 757.

129. Yamanaka, C., Mima, K. Nakai, S., Yamanaka, T., Izawa, Y., Kato, Y., *et al.*, (1987 a), *Inertial confinement fusion research by GEKKO lasers at ILE Osaka and target design for ignition.* in: Plasma Phys. and Controll, Nucl. Fusion Res. (IAEA, Vienna, 1986), **3**, 33.

130. Yamanaka, C., Nakai, S., Yamanaka, T., Izawa, Y., Kato, Y., Mima, K., *et al.*, (1987 b), *High thermonuclear neutron yield by shock multiplexing implosion with GEKKO XII green laser.* Nuclear Fusion, **27**, 19.

131. Zel'dovich, Ya. B., and Raizer, Yu. P., (1966), *Physics of Shock Waves and High-temperature Hydrodynamic Phenomena* (Academic, New York).

INDEX

LASER SCIENCE AND TECHNOLOGY
An International Handbook

SECTIONS

Distributed Feedback Lasers

Laser Fusion

Lasers and Surfaces

Laser Photochemistry

Chaos and Laser Instabilities

Excimer Lasers

Lasers and Nuclear Physics

Lasers in Medicine and Biology

Lasers and Fundamental Physics

Laser Spectral Analysis

Lasers and Communication

Solid State Lasers

Laser Diagnostics in Chemistry

Semiconductor Diode Lasers

Topics in Theoretical Quantum Optics

Gas Lasers

Fiber Optics Devices

Ultrashort Pulses and Applications

Coherent Phenomena

Coherent Sources for VUV-Radiation

Optical Storage and Memory

Tunable Lasers for Spectroscopy

Topics in Nonlinear Optics

Frequency Stable Lasers and Applications

Interaction of Laser Light with Matter

Mechanical Action of Laser Light

New Solid State Lasers

Laser Monitoring of the Atmosphere

Optical Bistability

Vapour Deposition of Materials on Surfaces

Optical Computers

Phase Conjugation

Phase Spectroscopy

Multiphoton Ionization of Atoms and Molecules

Squeezed States of Light

PUBLISHED TITLES

Further titles in preparation